Learning Over Time: Learning Trajectories in Mathematics Education

Learning Over Time: Learning Trajectories in Mathematics Education

Edited By

Alan P. Maloney
North Carolina State University

Jere Confrey
North Carolina State University

Kenny H. Nguyen
Catlin Gabel School

INFORMATION AGE PUBLISHING, INC.
Charlotte, NC • www.infoagepub.com

Library of Congress Cataloging-in-Publication Data

Learning over time : learning trajectories in mathematics education / edited by Alan P. Maloney, Friday Institute for Educational Innovation, Jere Confrey, Friday Institute for Educational Innovation and Amplify Learning, Kenny H. Nguyen, Catlin Gabel School.

pages cm

ISBN 978-1-62396-568-6 (pbk.) – ISBN 978-1-62396-569-3 (hardcover) – ISBN 978-1-62396-570-9 (ebook) 1. Mathematics–Study and teaching (Early childhood) 2. Educational psychology. 3. Child development. I. Maloney, Alan P., editor of compil ation. II. Confrey, Jere, editor of compilation. III. Nguyen, Kenny H., editor of compilation.

QA135.6.L434 2014

372.7–dc 3

2013048596

Printed in the United States of America

CONTENTS

PREFACE

Richard A. Duschl

The growth and development of educational researchers' ideas and understandings about cognition and learning and about the structure of disciplinary knowledge and professional practices has led to our recognizing the importance of (1) sequence and coherence in the design of curriculum, instruction, and assessment and (2) instruction-assisted development. Decades of thinking and research in what I refer to as the three "P" frameworks in education—philosophy, psychology, and pedagogy—have aligned to provide us with new views of learning, of adaptive instruction that aids learning, and of the design of assessment-driven learning environments that keep learning moving forward.

Since 2000, there have been several U.S. National Research Council (NRC) synthesis reports on learning, assessing learning, mathematics learning, and science learning. A policy and practice outcome of the NRC reports is the development of the new Core Common State Standards for Mathematics and the Next Generation Science Standards. Here one sees the commitment but also the challenge for curriculum and assessment to be tightly configured as well as aligned with instruction. What we teach, how we teach, and when we begin to teach a topic are all being reconsidered.

Learning Over Time: Learning Trajectories in Mathematics Education, pages vii–ix.

The push for practices and core ideas in mathematics and science education has some researchers thinking hard about the common ground between K–16 science and math curricula. Researchers from each of these disciplinary domains also have begun to focus on developmental pathways or corridors that inform the sequence and coherence of instruction, within a grade level as well as across grade bands (e.g., Grades 3–5). How, for example, do learners' nascent capacities with quantitative reasoning and modeling begin? Some would argue they begin with the development of mathematical ideas and science practices embedded in measurement, data and data displays, statistics and inference, and probability. Mapping out these and other sequences is the goal for the emergent research of *learning trajectories* in mathematics and of *learning progressions* in science.

The volume you hold contains several fine contributions of fundamental mathematics education research that is developing our understandings of learning trajectories. There are also contributions that examine how the use or extension of learning trajectories may (and should) influence the development of textbooks, curriculum analysis, professional development, assessments, and standards. The contributors are leading scholars at the forefront of cutting-edge perspectives regarding thinking about learning trajectories in mathematics as well as learning progressions in the sciences.

The editors and authors came together in summer 2009 to attend a conference held at the Friday Institute for Educational Innovation, North Carolina State University. Many of the chapters in this volume are derived from papers presented at that conference. A year later I caught up with this conversation at the fall 2010 conference, Designing Technology-Enabled Diagnostic Assessments for K–12 Mathematics. Building on the 2009 papers, I found the presentations and thematic breakout sessions to be thoughtful, insightful, and stimulating. I went to the 2010 conference wondering what I would learn and came away energized with a realization that we are much farther down the road than I was aware with respect to the format of learning trajectories and role of diagnostic assessments. Reading the papers from the 2009 conference further enriched my perspective and my excitement about learning trajectory and learning progression research and development.

While there is certainly a vision among the volume's authors as well as many practical examples of how learning trajectories can positively influence student learning and teacher professional development, there is, of course, much work still to be done. Designing curriculum and instruction that attend to sequence, coherence, trajectories, and progressions requires concomitant development of assessment tools, tasks, and techniques for measuring and monitoring learners' movement along these trajectories and progressions. One cannot underestimate the challenges as well as the robust opportunities that lie ahead for teachers and learners in K–12 education and in the postsecondary STEM disciplines.

In science and mathematics over the last century, we have learned how to learn about nature. In education over the last half century, we have learned how to learn about learning. As we proceed into the 21st century, we need to learn how to meld together these two endeavors. The contributions found in this edited volume represent thoughtful and insightful examinations of the curricular, methodological, pedagogical, and professional development challenges we face in the pursuit of ratcheting up STEM education. There are messages here certainly for mathematics education, but this is a valuable volume for science education, too. So I invite you to become energized about—and join me in the conversations regarding—the design of technologies, tools, and theories that inform developing learning pathways and building diagnostic assessments that move mathematics, science, and STEM learning forward with learning trajectories and learning progressions.

INTRODUCTION

LEARNING TRAJECTORIES IN MATHEMATICS

Jere Confrey, Alan P. Maloney, and Kenny H. Nguyen

SETTING THE STAGE FOR LEARNING TRAJECTORIES

Currently, discussion of *learning trajectories* (or *learning progressions*) is nearly ubiquitous within the mathematics education and science education communities in the United States. The learning trajectory is an idea that has come to the fore because of its potential to support new means of assessing the evolution of children's thinking about "big ideas" over time (Duschl, Schweingruber, & Shouse, 2007; Reyna, Chapman, Dougherty, & Confrey, 2011) and its role as an organizer for Common Core State Standards (National Governors Association Center for Best Practices & Council of Chief State School Officers, 2010). Concerned with the development of big ideas over longer periods of time, needing to provide teachers better means to support gradual learning over time, and dissatisfied with the quality of student thinking as measured by multiple-choice assessments, researchers have proposed that learning trajectories can address these educational needs. As stated in *Taking Science to School,*

Learning Over Time: Learning Trajectories in Mathematics Education, pages xi–xxii.

> A challenge is to understand the pathways—or learning progressions—by which children can bridge their starting point and the desired end point. Given the complexity and counterintuitive nature of the end point, such learning must necessarily occur over a long period of time, work on multiple fronts, and require explicit instruction. Yet at present, curriculum sequences are not typically guided by such long-term vision or understanding, nor is there clear agreement, given the wealth of scientific knowledge, about what might be truly foundational and most important to teach. (Duschl et al., 2007, p. 214)

In education, it is risky when an idea gains too much attention too quickly, especially one that is as complex as the idea of learning trajectories, notwithstanding that some of the work on learning trajectories by authors in this volume dates back a decade (Battista, 2004; Clements & Sarama, 2004). How does one avoid having learning trajectories devolve into yet another short-lived fad, viewed as a promised silver bullet to cure all woes, but which then collapses under the pressures of naïve expectations and impatience, excessive promises, and lack of careful and steady progress forward? The authors in this book understand that the goal of a learning trajectories approach is to establish the scientific foundations of students' learning of big ideas over time. That is, the goal is to extend and deploy what the fields of mathematics and science education have painstakingly learned about patterns in student reasoning that progress from prior knowledge, through intermediate states in developing understanding, and interwoven with the domain goals of learning—and to forge ways with teachers, as partners, to use this scientific foundation to systematically strengthen instruction.

This volume is an outgrowth of a national conference, cosponsored by the DELTA (Diagnostic E-Learning Trajectories Approach) Group, led by North Carolina State University (NCSU) professors Jere Confrey and Alan Maloney (two of the volume's editors), and the Center on Continuous Instructional Improvement (CCII), held at the Friday Institute for Educational Innovation at North Carolina State University in Raleigh, North Carolina, in late summer of 2009. Most of the authors in this book made presentations and led discussions at that conference. The conference was attended by numerous experts in mathematics education. Also attending were members of the writing team and the organizational sponsors of the Common Core State Standards (CCSS), who subsequently sought the advice of several of the authors, including organizer Confrey and contributing authors Battista, Clements, and Lehrer, during the writing and validation of the Common Core State Standards for Mathematics (CCSS-M). Many insights from the conference itself have been included in a previous report (Corcoran, Mosher, & Daro, 2011). The chapters in the present volume go into detail on specific learning trajectory research, and provide updated accounts of the relevant research and application projects.

The *hypothetical learning trajectory* was introduced in 1995 as a component of a "mathematics teaching cycle" developed in part from teaching experiments and grounded in constructivist perspectives on learning, as a tool for individual teachers to make sense of their own students' day-to-day progress and to frame their moment-to-moment and day-to-day instructional planning (Simon, 1995). By the 2010s, the learning trajectory (or learning progression) has become much more than a framework for an individual teacher's daily instructional planning, observation, and reflection. Anchored firmly to the work of instruction, *learning trajectory* in the second decade of the 21st century now increasingly signals research that aims for a systematic, detailed description of the likely progression of children's reasoning about big ideas of mathematics over long periods of time.

The authors of the present volume all grapple in one way or another with issues that have been at the heart of mathematics education for several decades:

- How students come to understand mathematical concepts and build increasingly sophisticated reasoning, and how this learning corresponds to experts' conceptions of the discipline of mathematics
- How to integrate an increasingly robust research base into classroom practice, curriculum analysis, and educational policy; to instructionally scaffold and leverage the experience of students from diverse points of entry into increased success in mathematics; and to catalyze stronger student engagement, agency, and intellectual growth
- How the use of learning trajectories in professional development can support teachers in deepening their own conceptual understanding and their ability to understand and nurture students' conceptual development
- How to develop nuanced understanding of children's progression through big ideas using formative assessment and counterbalance the pressure of high-stakes assessment and "teaching to the test."

The first section of this volume presents examples of research into the students' development of concepts and reasoning over a broad time frame. The learning trajectory construct is a three-part framework for describing the development over time of children's reasoning about big ideas in mathematics: (1) prior knowledge and the entry to reasoning about the mathematical concept under consideration; (2) task- and instruction-based progression through intermediate states of understanding, landmark concepts, and obstacles or misconceptions; and (3) a high-level conceptual objective or particular "goal understanding" within the domain. Researchers employ a suite of related methodologies (clinical interviews, design studies, and teaching experiments). The researchers recognize that the experiences that children bring to the classroom, and that the environment, tasks, and instruction children encounter in the classroom, all affect the ways students progress through intermediate understandings, as they develop increasingly sophisticated reasoning and periodically reorganize their understanding. The development of rich student cognition is inseparable from the interactions among students and teachers, in classrooms and other venues. There is a common thread among this body of research—namely, that characterizing children's reasoning will allow students to gain confidence as learners and will allow teachers to support children's development as increasingly sophisticated conceptual learners. How to represent learning trajectories and how to characterize the interactions among learning trajectories in instruction and curriculum are major challenges.

The chapters in the second section of this volume illustrate how learning trajectories are increasingly relied on to support coherence based on student learning across other components of education, including curriculum analysis and development, assessment, and—perhaps most significant for sustaining the use of learning trajectories in all these contexts—educational standards. One of the challenges in this work is to develop representations of learning trajectories that are useful in these distinct but highly interrelated contexts while sustaining the essential focus on the progression of student learning. Thus, several chapters in this section suggest implicitly or explicitly that the learning trajectory is proving to be a powerful boundary object that facilitates coherence across classroom instructional practice, professional development, instructional leadership, research, and the design, implementation, and revision of standards and assessments (Penuel, Confrey, Maloney, & Rupp, in press).

SECTION ONE: RESEARCH TO CONSTRUCT LEARNING TRAJECTORIES

The key idea behind learning trajectories is the depiction of how students progress toward increasingly sophisticated understanding and productive reasoning around one or more related concepts or big ideas in mathematics. Over the past 30 to 40 years, a broad variety of research has spoken to how students make progress toward understanding some of the key ideas, skills, and strategies in mathematics; the resulting literature relevant to individual big ideas, however, is typically fragmented due to the variations in research questions and methods. To a degree, the concept of learning trajectories helps to synthesize those different contributions.

The first four chapters in this volume address theoretical and empirical foundations of mathematics learning trajectories and provide detailed examples of relevant research. The various authors articulate differently nuanced definitions of learning trajectories. Clements and Sarama (Chapter 1) elucidate the progression of student cognition that leads to fundamental early mathematical concepts such as counting and number sense.

They review their own powerful theoretical framework (hierarchic interactionalism), of which learning trajectories are a core tenet. They provide a succinct recounting of how learning trajectories add to a long history of research, a major advance that distinguishes them from previous research in a number of important ways: The goals are based both on domain-specific expertise *and* research into children's thinking and learning; they incorporate a cognitive science view of learning as interconnected concepts and skills characterized by mental objects and actions; and instructional tasks are critical and must be carefully designed to engender the reasoning characteristic of the levels of thinking. Clements and Sarama have built an entire K–2 curriculum (*Building Blocks*) around their learning trajectories for early geometric and spatial ideas and skills and numeric and quantitative ideas and skills, all based in children's activities that lead to these concepts and skills. Multiple curricular evaluation studies demonstrated a large and statistically significant effect size associated with the use of their curriculum compared with more traditional curricula. Their work stands as an example of a long-term research program, one based on the study of student learning, that has produced important findings on improved student learning, practical products for teachers' use, and foundations for further research.

Lehrer and colleagues (Chapter 2) have established and validated a learning progression focused on student and teacher exploration of data, measurement, and the understanding of variability, topics that underlie student reasoning about statistics and mathematical and scientific modeling. They remind us that "establishing a learning progression is an epistemic enterprise" of its own. Their extensive research involving design study and evidence-centered design approaches to assessment demonstrate the remarkable depth and sophistication of reasoning students can develop in these topics in late elementary and early middle grades, when a data-modeling perspective is used instead of the typical piecemeal approach to statistical reasoning. Unlike many authors, Lehrer et al. distinguish *learning trajectory* from *learning progression*. From their perspective, a learning progression is the broad repertoire of means "for nurturing the long-term development of disciplinary knowledge and dispositions" (see Chapter 2, "Initial Considerations for Designing a Learning Progression") and broadly encompasses professional development, assessment, and institutional organization, whereas the learning trajectory itself is more limited in scope, a conjecture about idealized states of knowledge (paraphrasing Simon, 1995).

Confrey and colleagues (Chapter 3) have validated a previously conjectured mathematical construct, equipartitioning, that undergirds the early development of rational number reasoning. In developing the equipartitioning learning trajectory, Confrey et al. synthesized various threads of research focused on topics that had individually been seen in curriculum and instruction—one-to-one correspondence, early understanding of fraction, ratio, and measurement—but which taken together embody a more inclusive cognitive construct. The authors identified equipartitioning as a distinct construct and developed proficiency levels for the trajectory, from young children's sharing of collections to the generalization that *m* objects shared among *p* people results in *m*/*p*th objects per person. Their analysis came from combining prior theory on "splitting" (Confrey, 1989) with a synthesis across three cases: sharing a collection, sharing a single whole, and sharing multiple wholes. The authors further link the proficiency levels to task classes in order to view how changes in shape or splitting number interact with the proficiency levels.

Barrett and Battista (Chapter 4) compare two different learning trajectories for shape and measurement, along with different representations for those trajectories, that have been developed by Battista (Battista & Clements, 1996; Battista, Clements, Arnoff, Battista, & Borrow, 1998) and by Clements, Sarama, and Barrett (Barrett & Clements, 2003; Clements & Sarama, 2009). Their chapter makes it clear that while different trajectories will have some significant commonalities, they may differ in how they view the meaning of the levels, in their specific relationship to curriculum and assessments, and in the particular aspects of the trajectory that are emphasized. Comparing and making meaning between different trajectories in the same domain will no doubt be an important task for the field as research on learning trajectories expands.

The researchers who develop learning trajectories, and those who use them to better understand student learning, assume that children must be taken seriously as learners—people who are trying to make sense of the world and are continually taking on new intellectual challenges. Learning trajectory research aims to systematically document children's reasoning through their inscriptions; their utterances; and exemplars of their interactions with materials, ideas, each other, and instruction. Mathematics, cognitive psychology, student-centered learning and research, and constructivism are all drawn on and woven together in the learning trajectory framework.

The learning trajectories discussed in this volume reflect extensive empirical research (clinical interviews, teaching experiments, and field tests, as well as curriculum development), investigating conjectures of the most likely tendencies of students' progress toward domain goal understandings through carefully designed tasks and instruction. When one recognizes that learning trajectories incorporate an "envelope" of expected proficiencies for any roomful of students, and provide the basis for scaffolding the conceptual development of individual children, the potential value of learning trajectories to teachers is readily apparent. They provide an underlying theory of how students learn mathematics, against which teachers' interactions with their students, and students' learning itself, can be calibrated.

We see several lessons echoed in each of the research-based chapters. These include:

1. Learning trajectories are empirically supported descriptions of the likely obstacles and landmarks students encounter as they move from naïve to more sophisticated ideas. They are not a set of thought experiments, or logically derived predictions about how students might learn.
2. Learning trajectories differ from propositional knowledge. In learning trajectories, partial ideas may serve to generate the concept, because partial ideas that may later prove to be naïve, incomplete, or "wrong" often play critical roles in the genesis of foundational ideas and robust reasoning. From the learner's perspective, these partial ideas help solve a local challenge or a problematic through encountering and solving tasks, and thus satisfy a genetic need as students progress towards the more sophisticated positions. These intermediate proficiencies are characterized by what they facilitate in a child—and in his or her interactions with other learners and the instructor. Learning trajectories provide a language and framework to describe and signify these intermediate proficiencies and alert the teacher to how they develop over time into more robust mathematical properties and generalizations.
3. A learning trajectories approach is not a stage theory, even though typically represented as a sequence (in part due to current constraints in representations employed to describe and study learning trajectories). The sequence acts as a guide to what students are *likely* to say and do and shows how those proclivities provide evidence of particular underlying beliefs and understandings. Learning trajectories represent expected probabilities of student reasoning within the context of instructional experiences. They do not necessarily represent a series of developmental prerequisites. Misrepresenting them as a stage theory would constrain the pace and variability of learning and unintentionally deny students opportunities to generate, test, and express their own ideas.
4. Learning trajectories depend on a perspective of *genetic epistemology* (Piaget, 1970), which recognizes that mathematical reasoning is essential to the development of mathematical concepts. This does not mean simply valuing both procedural and conceptual knowledge (National Research Council, 2001), but rather the notion that *how* one (a student) comes to believe an idea determines the nature, quality, and viability of that idea. It is necessary to tie the idea to be learned to the problematic it addresses—that is, to create and fulfill a cognitive need. The content and the practices of

doing mathematics, then, provide the learner with an understanding of the explanatory power in the ideas.

5. As learning trajectories are informed by Piaget's investigations, they are also informed by a sociocultural framework (Vygotsky, 1986). In short, instruction matters. Learning is scaffolded by careful sequencing of activities and tasks and by fostering discourse and interactions. In addition, what is learned is profoundly affected by the selection of tools as mediating features. Learning trajectories depend on instruction, through which students are provided relevant and carefully sequenced tasks or projects, a rich variety of tools and resources available, and opportunities for discourse and discussion.
6. Effective use of learning trajectories depends on teachers learning to trust and support the emergence of student ideas as central to the flow, process, and act of instruction. It relies on the belief that teaching is accomplished not by "simply" replacing children's thinking with adult thinking, but rather by facilitating the transformation of children's thinking towards sustained participation, problem solving, and rich understanding.
7. Uniformity in student prior knowledge, learning, or progress cannot be expected. Individual students are expected to demonstrate variability in their progress due to particular past experiences, language proficiency, tool use, perceptions, or culture. A critical element in learning trajectory research (and its use in instruction) is to understand and support the diversity of student ideas while seeking predictable patterns of student reasoning and behavior that experience tells us will emerge.
8. Multiple learning trajectories can exist for similar domain goal understandings. Even as the lessons from learning trajectories research are the outcome of scientific research, the lessons are not the educational equivalents of physical laws. They express the patterns of predictable results that accrue under certain sequences of tasks, availability, and use of tools and the elaboration of discourse and instructional moves.

SECTION TWO: APPLICATIONS AND IMPLICATIONS OF LEARNING TRAJECTORIES

In the second section, a new set of insights that emerges from these authors' research is how these learning trajectories can be deployed at multiple levels of the educational system to promote coherence among different levels of organization in mathematics education.

Many of the researchers attending the 2009 conference, among others, advocated for the use of learning trajectories in the development of the

Common Core State Standards for Mathematics (National Governors Association Center for Best Practices & Council of Chief State School Officers, 2010); the CCSS-M's emphasis on learning progressions has become widely recognized as a strength of the new standards. However, the standards document lists the standards by grade level, and it can be problematic to readily visualize the continuity of one topic or standard to the next within or between grades. In Chapter 5, Confrey and Maloney discuss how a learning trajectories perspective provides a productive underpinning for new representations of the Common Core Standards. Displays such as those featured in Chapter 5 provide valuable visual and textual pathways linking the big ideas that must be fostered within and across the grades. Chapter 5 also broadens the notion of learning trajectories as boundary objects, first raised by Lehrer et al. in Chapter 2. As boundary objects, mediating communication, and shared vision across the boundaries of different spheres of influence in the educational system, they can support an underlying framework for instruction, professional development, curricula design, and classroom assessment. Within the activity at the instructional core (Elmore, 2002), learning trajectories can mediate cohesion to curriculum, instructional practice, classroom discourse, and professional development. A challenge in research *and* policy environments is to generate a broad enough base of explicit learning trajectory research, including professional development experience, and to put in place a model for revising the Common Core standards over time by incorporating new learning trajectory research.

Nguyen and Confrey (Chapter 6) take on the challenge of distinguishing the differences between learning trajectories and the written curriculum. They argue that the two, while sharing some similarities, are inherently different. While the written curriculum is static, learning trajectories provide a theory that, when used as a lens through which to view the written curriculum, has the potential to provide flexibility in teaching. To support this, they performed a content analysis, from a learning trajectories perspective, of rational number reasoning within two middle school curricula. This analysis reveals the extent to which each curriculum aligns with extant rational number reasoning trajectories, and the extent to which gaps occur in rational number reasoning coverage in each. The authors conclude that understanding learning trajectories is vital because it allows teachers to modify their written curriculum in order to ensure that the enacted curriculum is aligned with best practices from research. They suggest that performing content analyses from a learning trajectories perspective for both curriculum and standards (e.g., CCSS-M) is an important endeavor for both teachers and researchers moving forward in this era of reform.

Olson (Chapter 7) describes a framework for discerning learning trajectories for two algebraic constructs, inasmuch as they are articulated in middle-school mathematics textbooks. He discusses findings through the use of a framework, articulated learning trajectories (ALT), to analyze textbooks. An ALT is a way to delineate "a precise written sequence of mathematics content that authors of textbooks envision teachers and students will follow or use to guide the development of mathematics content within the classroom…a baseline of how a concept develops in written text" (see Chapter 7, Articulated Learning Trajectories). Among four textbook series, Olson finds considerable variation in the number of instances involving patterning constructs and contexts, the textual placement and emphasis of particular patterning constructs and contexts, and in the development of algebraic thinking concepts as defined by the ALTs he identified within and across the textbooks and textbook series.

Wilson (Chapter 8) illustrates the use of the equipartitioning learning trajectory developed by Confrey et al. (Chapter 3) to support growth in a coherent understanding by elementary level teachers of this early rational number reasoning construct and in supporting their growth in predicting and interpreting student behaviors and responses. The empirically developed learning trajectory was used in professional development as a framework shared across many classrooms and multiple schools. By linking learning trajectories to teacher education and professional development, teacher educators are devising ways to make this information available to and understood by teachers.

Lessons we see emerging from the application-based chapters include the following:

1. The application of a learning trajectories-based analysis of standards can reveal whether standards successfully carry the development of concepts over time while coordinating the interrelationships among learning trajectories.
2. Proficiency levels represent intermediate states of learning and emphasize the importance of gradual and constant growth in student thinking. When called up by standards, those states may also be viewed as targets for assessment. This results in a tension between the articulation of standards as learning goals and their use in identifying content that is assessable.
3. In constructivist-oriented or student-centered classrooms, learning trajectories can assist in the construction of rich curriculum and activities and in the choice of tools and materials. The learning trajectories can affect the design of the curriculum by improving curricular coherence, supporting the emergence of student ideas, and fostering conceptual growth.
4. A strong familiarity with learning trajectories, because they are developed from empirical study of student patterns of understanding and careful design of tasks to support conceptual development,

> can help teachers recognize the implications of student questions, conjectures, errors and misconceptions, and use of language. They can provide teachers, both practicing and prospective, with a systematic way to learn to recognize the possibilities and inherent logic in student thinking. They also help them to recognize sticky misconceptions and find paths to recognize and address them appropriately.

As you explore this volume, we hope that you see how the authors share a framework for research in student learning that richly extends our understanding and characterization of how students learn mathematics. The authors featured in this volume are committed to applying this research to transform mathematics instruction, through the research enterprise itself, to development of curriculum, assessment, and professional development that embodies the learning trajectory construct. Learning trajectories can lend a coherence to mathematics learning and teaching that crosses boundaries of usually distinct compartments of the instructional mission: from the daily episodes of teaching throughout the instructional core—the curriculum, classroom discourse, and formative and diagnostic assessment—to the bookends of the accountability system—standards and policy development and high stakes assessment.

REFERENCES

Barrett, J. E., & Clements, D. H. (2003). Quantifying path length: Fourth-grade children's developing abstractions for linear measurement. *Cognition and Instruction, 21*(4), 475–520.

Battista, M. T. (2004). Applying cognition-based assessment to elementary school students' development of understanding of area and volume measurement. *Mathematical Thinking and Learning, 6*(2), 185–204.

Battista, M. T., & Clements, D. H. (1996). Students' understanding of three-dimensional rectangular arrays of cubes. *Journal for Research in Mathematics Education, 27*(3), 258–292.

Battista, M. T., Clements, D. H., Arnoff, J., Battista, K., & Borrow, C. V. (1998). Students' spatial structuring of 2d arrays of squares. *Journal for Research in Mathematics Education, 29*(5), 503–532.

Clements, D. H., & Sarama, J. (2004). Learning trajectories in mathematics education. *Mathematical Thinking and Learning, 6*(2), 81–89.

Clements, D. H., & Sarama, J. (2009). *Learning and teaching early math: The learning trajectories approach.* New York, NY: Routledge.

Confrey, J. (1988). Multiplication and splitting: Their role in understanding exponential functions. In M. Behr, C. Lacampagne, & M. M. Wheeler, (Eds.), *Proceedings of the Tenth Annual Meeting of the North American Chapter of the International Group for the Psychology of Mathematics Education* (pp. 250–259). Dekalb, IL: Northern Illinois University.

Corcoran, T., Mosher, F. A., & Daro, P. (Eds.). (2011). *Learning trajectories and progressions in mathematics.* New York, NY: Consortium for Policy Research in Education.

Duschl, R. A., Schweingruber, H. A., & Shouse, A. W. (Eds.). (2007). *Taking science to school: Learning and teaching science in grades K-8.* Washington, DC: National Academies Press.

Elmore, R. F. (2002). *Bridging the gap between standards and achievement.* Washington, DC: Albert Shanker Institute.

National Governors Association Center for Best Practices & Council of Chief State School Officers. (2010). *Common core state standards for mathematics.* Washington, DC: Author.

National Research Council. (2001). *Adding it up: Helping children learn mathematics.* Washington, DC: National Academy Press.

Penuel, W. R., Confrey, J., Maloney, A. P., & Rupp, A. A. (in press). Design decisions in developing assessments of learning trajectories: A case study. *Journal of the Learning Sciences.*

Piaget, J. (1970). *Genetic epistemology.* New York, NY: W.W. Norton.

Reyna, V. F., Chapman, S. B., Dougherty, M. R., & Confrey, J. (Eds.). (2011). *The adolescent brain: Learning, reasoning, and decision making.* Washington, DC: American Psychological Association.

Simon, M. A. (1995). Reconstructing mathematics pedagogy from a constructivist perspective. *Journal for Research in Mathematics Education, 26*(2), 114–145.

Vygotsky, L. S. (1986). *Thought and language* (A. Kozulin, Trans.). Cambridge, MA: MIT Press.

CHAPTER 1

LEARNING TRAJECTORIES

Foundations for Effective, Research-based Education

Douglas H. Clements and Julie Sarama

ABSTRACT

Approaches to standards, curriculum development, and pedagogy are remarkably diverse; however, recent years have seen a growing movement to base each of these on learning trajectories. In this chapter, we discuss and compare the various terms and conceptions of this construct, present our own definition, differentiate between our conception and that of others, and briefly review the current evidentiary base in our area, early childhood mathematics.

THE LEARNING TRAJECTORY CONSTRUCT

Learning trajectories are a device whose purpose is to support the development of a curriculum, or a curriculum component. The term *curriculum* stems from the Latin word for race course, referring to the course of expe-

Learning Over Time: Learning Trajectories in Mathematics Education, pages 1–30.

riences through which students grow to become mature adults. Thus, the notion of a path, or trajectory, has always been central to curriculum development and study. In his seminal work, Simon stated that a *hypothetical learning trajectory* included "the learning goal, the learning activities, and the thinking and learning in which the students might engage" (1995, p. 133).

Building on Simon's definition, but emphasizing a cognitive science perspective and a base of empirical research,

> We conceptualize learning trajectories as descriptions of children's thinking and learning in a specific mathematical domain, and a related, conjectured route through a set of instructional tasks designed to engender those mental processes or actions hypothesized to move children through a developmental progression of levels of thinking, created with the intent of supporting children's achievement of specific goals in that mathematical domain. (Clements & Sarama, 2004b, p. 83)

In other words, each learning trajectory has three parts: (1) a goal, (2) a developmental progression, and (3) instructional activities. To attain a certain mathematical competence in a given topic or domain (the goal), students learn each successive level (the developmental progression), aided by tasks (instructional activities) designed to build the mental actions-on-objects that enable thinking at each higher level (Clements & Sarama, 2004).

The term *learning trajectory* reflects its roots in a constructivist perspective. That is, although the name emphasizes learning over teaching, both the Simon and the Clements and Sarama definitions above clearly involve teaching and instructional tasks. Some interpretations and appropriations of the learning trajectory construct emphasize only the developmental progressions of learning during the creation of a particular curricular or pedagogical context. Some terms, such as *learning progressions,* are used ambiguously, sometimes indicating developmental progressions, and at other times suggesting a sequence of instructional activities. Although studying either psychological developmental progressions or instructional sequences separately can be valid research goals, and studies of each can and should inform mathematics education, we believe the power and uniqueness of the learning trajectory construct stems from the inextricable interconnection between these two aspects.

Both these aspects (developmental progressions of thinking and instructional sequences) serve the most important, but often least discussed, aspect of learning trajectories—the goal. The goals of our learning trajectories are derived from the expertise of mathematicians as well as research on students' thinking about and learning of mathematics (Clements, Sarama, & DiBiase, 2004; Fuson, 2004; Sarama & Clements, 2009a). This results in goals that are organized into the "big," or focal, ideas of mathematics: overarching clusters and concepts and skills that are mathematically central and coherent, consistent with students' (often intuitive) thinking, and generative of future learning (Clements, Sarama, & DiBiase, 2004; National Council of Teachers of Mathematics, 2006).

Once the mathematical goals are established, research is reviewed to determine if there is a natural developmental progression (at least for a given age range of students in a particular culture) identified in theoretically and empirically grounded models of students' thinking, learning, and development (Carpenter & Moser, 1984; Griffin & Case, 1997). That is, researchers build a cognitive model of students' learning that is sufficiently explicit to describe the processes involved in the students' realization of the mathematical goal across several qualitatively distinct structural levels of increasing sophistication, complexity, abstraction, power, and generality. We will return to the methodologies used in a succeeding section.

The issue of what is meant by a "natural" developmental progression is sure to arise. We briefly present our theoretical framework, hierarchic interactionalism, because learning trajectories are its core tenet and because the other tenets help describe and justify the learning trajectories construct (for a full explication, see Sarama & Clements, 2009a). The term hierarchic interactionalism indicates the influence and interaction of global and local (domain-specific) cognitive levels and the interactions of innate competencies, internal resources, and experience (e.g., cultural tools and teaching). Students first represent mathematical ideas intuitively, then with language, then metacognitively, with the last indicating that the student possesses an understanding of the topic and can access and operate on those understandings. The tenets of hierarchic interactionalism include the following, which are relevant to the learning trajectories construct:

1. *Developmental progression.* Most content knowledge is acquired along developmental progressions of levels of thinking. These progressions play a special role in students' cognition and learning because they are particularly consistent with students' intuitive knowledge and patterns of thinking and learning at various levels of development.
2. *Domain-specific progression.* These developmental progressions often are most propitiously characterized within a specific mathematical domain or topic (Dowker, 2005; Karmiloff-Smith, 1992; Resnick, 1994; Van de Rijt & Van Luit, 1999). Students' knowledge—that is, the objects and actions they have developed in that domain—are the main determinant of the thinking within each progression, although hierarchic interactions occur at multiple levels within and between topics, as well as with general cognitive processes (e.g., executive, or metacognitive, processes; potentialities for general rea-

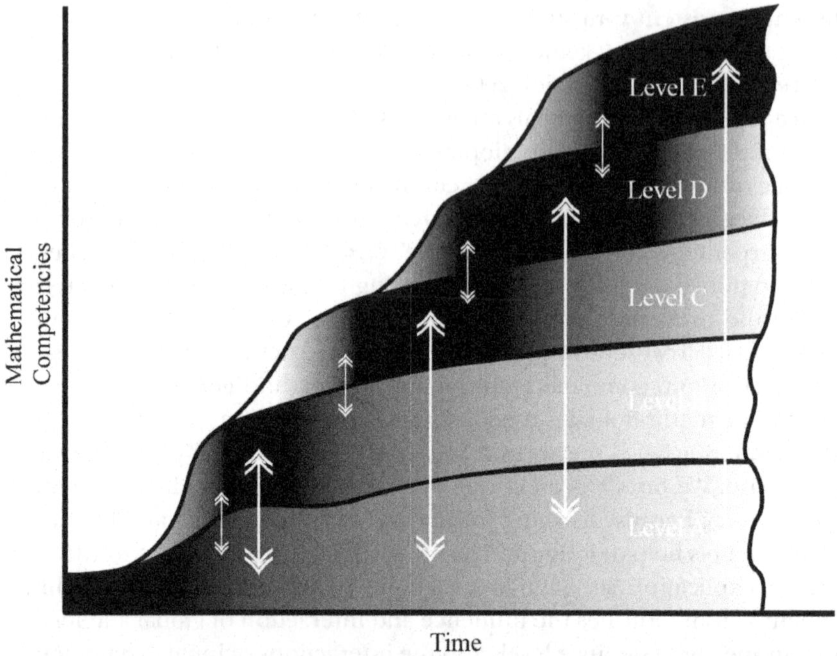

Note: In the Hierarchic Interactionalism view illustrated here, types of knowledge develop simultaneously. Darker shading indicates dominance of a particular level of thinking at a particular point in time. Thus, Level A is dominant in the earliest years, but Level B thinking begins to develop and eventually interacts with Level A thinking, although only weakly at first (symbolized by the small double arrow at the left). Once Level B thinking is more developed and more strongly connected with Level A, it becomes the dominant pattern of thinking. Shading indicates the unconscious probabilities of instantiation; in a related vein, but not illustrated, are the executive processes that also develop over time, serving to integrate these types of reasoning and, importantly, to determine which level of reasoning will be applied to a particular situation or task.

FIGURE 1.1. Developmental progression with hierarchic interactionalism
Note: Idealized view of levels of thinking in a learning trajectory's developmental progression, as hypothesized in hierarchic interactionalism (adapted from Sarama & Clements, 2009a). In the hierarchic interactionalism view illustrated here, types of knowledge develop simultaneously. Darker shading indicates dominance of a particular level of thinking at a particular point in time. Thus, Level A is dominant in the earliest years, but Level B thinking begins to develop and eventually interacts with Level A thinking, although only weakly at first (symbolized by the small double arrow at the left). Once Level B thinking is more developed and more strongly connected with Level A, it becomes the dominant pattern of thinking. Shading indicates the unconscious probabilities of instantiation; in a related vein, but not illustrated, are the executive processes that also develop over time, serving to integrate these types of reasoning, and, importantly, to determine which level of reasoning will be applied to a particular situation or task.

soning and learning-to-learn skills; and some other domain-general developmental processes). See Figure 1.1 for an illustration.

3. *Hierarchic development.* Development is less about the emergence of entirely new processes and products and more about an interactive

interplay among specific existing components of knowledge and processes. Also, each level builds hierarchically on the concepts and processes of the previous levels. Developmental progressions within a domain can repeat themselves in new contexts (Siegler & Booth, 2004). The learning process is more often incremental and gradually integrative than intermittent and tumultuous. Various models and types of thinking grow in tandem to a degree, but a critical mass of ideas from each level must be constructed before thinking characteristic of the subsequent level becomes ascendant in the student's thinking and behavior (Clements, Battista, & Sarama, 2001). Successful application leads to the increasing use of a particular level. However, under conditions of increased task complexity, stress, or failure this probability level decreases and an earlier level serves as a fallback position (Hershkowitz & Dreyfus, 1991; Siegler & Alibali, 2005).

4. *Co-mutual development of concepts and skills.* Concepts constrain procedures, and concepts and skills develop in constant interaction (Baroody, Lai, & Mix, 2005; Greeno, Riley, & Gelman, 1984). In imbalanced cultural or educational environments, one of the two may take precedence, to an extent that can be harmful. Effective instruction often places initial priority on conceptual understanding, including students' creations of solution procedures (Carpenter, Franke, Jacobs, Fennema, & Empson, 1998) and mathematical dispositions and beliefs: "Once students have learned procedures without understanding, it can be difficult to get them to engage in activities to help them understand the reasons underlying the procedure" (Kilpatrick, Swafford, & Findell, 2001, p. 122).
5. *Initial bootstraps.* Students have important, but often inchoate, premathematical and general cognitive competencies and predispositions at birth or soon thereafter that support and constrain, but do not absolutely direct, subsequent development of mathematics knowledge. Some of these have been called *experience-expectant processes* (Greenough, Black, & Wallace, 1987), in which universal experiences lead to an interaction of inborn capabilities and environmental inputs that guide development in similar ways across cultures and individuals. They are not built-in representations or knowledge, but predispositions and pathways to guide the development of knowledge (cf. Karmiloff-Smith, 1992). Other general cognitive and metacognitive competencies make children—from birth—active participants in their learning and development (Tyler & McKenzie, 1990).
6. *Different developmental courses.* Different developmental courses are possible within those constraints, depending on individual, envi-

ronmental, and social confluences (Clements et al., 2001; Confrey & Kazak, 2006). Within any developmental course, at each level of development, students have a variety of cognitive tools—concepts, strategies, skills, and utilization and situation knowledge—that coexist. The differences within and across individuals create variation that is the wellspring of invention and development. At a group level, however, these variations are not so wide as to vitiate the theoretical or practical usefulness of the tenet of developmental progressions; for example, in a class of 30, there may be only a handful of different solution strategies (Murata & Fuson, 2006), many of which represent different levels along the developmental progression.

7. *Progressive hierarchization.* Within and across developmental progressions, students gradually make connections between various mathematically relevant concepts and procedures, weaving ever more robust understandings that are hierarchical, in that they employ generalizations while maintaining differentiations. These generalizations and metacognitive abilities eventually connect to form logical-mathematical structures that virtually compel students toward decisions in certain domains, such as those on traditional Piagetian conservation of number tasks, that are resistant to confounding via misleading perceptual cues. Students provided with high-quality educational experiences build similar structures across a wide variety of mathematical domains. (Maintaining such hierarchical cognitive structures makes more demands on the educational environment and the learner as mathematics becomes more complex, beyond the early childhood years, a point to which we shall return.)
8. *Consistency of developmental progressions and instruction.* Instruction based on learning consistent with natural developmental progressions is more effective, efficient, and generative for the student than learning that does not follow these paths.
9. *Learning trajectories.* An implication of the tenets to this point is that a particularly fruitful instructional approach is one based on hypothetical learning trajectories (Clements & Sarama, 2004b). Beginning from specific, mental constructions (mental actions-on-objects) and patterns of thinking that are hypothesized to constitute students' thinking, curriculum developers design instructional tasks that include external objects and actions that mirror this hypothesized mathematical activity of students as closely as possible. These tasks are sequenced, each corresponding to a level of the developmental progressions, to complete the hypothesized learning trajectory. Such tasks will theoretically constitute a particularly efficacious educational program; however, there is no implication that the task sequence is the only path for learning and teaching,

only that it is hypothesized to be one fecund route. Tasks present a problem; people's actions and strategies to solve the problem are represented and discussed, and reflection on whether the problem is solved, or partially solved, leads to new understandings (mental actions and objects, organized into strategies and structures) and actions. Specific learning trajectories are the main bridge that connects the "grand theory" of hierarchic interactionalism to particular theories and educational practice (Confrey & Kazak, 2006).

10. *Instantiation of hypothetical learning trajectories.* Hypothetical learning trajectories must be interpreted by teachers and are only realized through the social interaction of teachers and students around instructional tasks:

> There is no understanding without reflection, and reflection is an activity students have to carry out themselves. No one else can do it for them. Yet a teacher who has some inkling as to where a particular student is in his or her conceptual development has a better chance of fostering a further reflective abstraction than one who merely follows the sequence of a preestablished curriculum. (von Glasersfeld, 1995, p. 382)

This theoretical framework establishes the basis for the instructional tasks—the third and final component of a learning trajectory. In our work, these are composed of key tasks designed to promote learning at a particular conceptual level or benchmark in the developmental progression. Extant research is used to identify tasks as effective in promoting the learning of students at each level, by encouraging students to construct the concepts and skills that characterize the succeeding level. That is, we hypothesize the specific mental constructions—the building of mental actions-on-objects and patterns of thinking that constitute students' thinking at each level. We design tasks that include external objects and actions that mirror the hypothesized mathematical activity of students as closely as possible. For example, objects may be shapes or sticks, and actions might be creating, copying, uniting, disembedding, and hiding both individual units and composite units. Tasks require students to apply, externally and mentally, the actions and objects of the goal level of thinking. (See Clements & Battista, 2000, for further description and examples of mathematical objects/concepts and mathematical actions/processes that operate on them.) These tasks are, of course, sequenced corresponding to the order of the developmental progressions to complete the hypothesized learning trajectory. Further, the characteristics of the tasks and their accompanying pedagogical interactions are explicitly linked to transitions between levels.

WHAT, IF ANYTHING, IS "NEW" IN THE LEARNING TRAJECTORIES CONSTRUCT?

When we discuss learning trajectories, some (commendably) skeptical colleagues ask what is really different (Lesh & Yoon, 2004). If curricula have always been courses, or paths (and frequently "horse races" along these), and if psychological and educational theories have always postulated series of goals, then is this not simply renaming old (and palpable) ideas?

At certain simple levels, the answer is: Yes, many of these notions share some characteristics. Most describe or dictate a series of educational goals. All have some theoretical perspective on why one goal might follow another. However, these theories often differ markedly in the details. The learning trajectory construct, as we define it, builds upon theories and research of years past, as any theory should, but is distinct from previous formulations and constitutes a substantive contribution to theory, empirical research, and praxis.

Early educational psychology considered series or educational sequences based on the accumulation of connections.

> We now understand that learning is essentially the formation of connections or bonds between situations and responses, that the satisfyingness of the result is the chief force that forms them, and that habit rules in the realm of thought as truly and as fully as in the realm of action. (Thorndike, 1922, p. v)

Thus, curricular sequences could be logically arranged to establish simple connections between simple situations (addends) and responses (sum), and then later connect these and other bonds to complete more difficult tasks (e.g., multidigit addition) and even to develop mathematical reasoning. However, conceptual, meaningful learning was not the focus, but rather simple, paired associate learning. Also, potential differences and nuances of learning in different subject matter domains was not considered.

Bloom's taxonomy of educational objectives and Robert Gagné's *conditions of learning* and *principles of instructional design* (Gagné, 1965; Gagné & Briggs, 1979) postulated that there were types of learning and that certain types, such as stimulus-response learning (e.g., Thorndike's *bonds*) were prerequisite to other types (e.g., verbal association, then concept learning, and last, problem solving). In a domain, they would be determined by logical analysis[1] and empirical task analysis.

Other researchers similarly based these hierarchies on logical and task analyses, but gave more weight to extant findings in educational and especially psychological research to perform cognitive, or rational, analyses (Resnick & Ford, 1981). Work from this perspective increasingly used computer metaphors (i.e., information-processing theories) and, often, computer models in their analyses (Hoz, 1979; Klahr & Wallace, 1976).

These approaches determined hierarchies of educational goals and were the basis of many "scope and sequences" in the educational literature. (See Baroody, Cibulskis, Lai, & Li, 2004, for an extended discussion and somewhat different perspective.) The view of learning in the earlier approaches was generally that of knowledge acquisition, with the environment providing input that was "received"—that is, imitated and mentally recorded by the student.

Later, researchers began to attend more to cognitive development. Some devised developmental learning theories in attempts to integrate structural views such as those of Piaget with views based on task analysis and information-processing models (Beilin, 1984). Later theoretical efforts in cognitive science extended these efforts to focus on the importance of domain-specific learning and development (Davis, 1984; Karmiloff-Smith, 1992).

In historical parallel, several theories, from Piagetian (Piaget & Szeminska, 1952) to field theories (Brownell, 1928; Brownell & Moser, 1949) and later developmental and cognitive science theories (Case & Okamoto, 1996), emphasized students as makers of meaning. Similarly, cognitively and constructivist-oriented research programs explicated the concepts and skills students build as they move from one level to the next within a mathematical domain (Baroody, 1987; Carpenter, Fennema, Franke, Levi, & Empson, 1999; Carpenter & Moser, 1984; Steffe & Cobb, 1988). Unfortunately, those seeking to apply these studies practically often oversimplified and misconstrued their results and implications, emphasizing an educational version of laissez-faire or outdated "discovery" approaches (Clements, 1997).

Learning trajectories, as we have defined them, owe much to these previous efforts, which have progressed to increasingly sophisticated and complex views of cognition and learning. However, the earliest applications of cognitive theory to educational sequences tended to feature simple linear sequences based on accretion of numerous facts and skills, as reflected in their hierarchies of educational goals and the resultant scope and sequences. Learning trajectories include hierarchies of goals and competencies but do not limit them solely to sequences of skills, as many of the earlier constructs did. Nor are learning trajectories lists of everything students need to learn; that is, they do not cover every single fact or skill. Further, they describe students' levels of thinking, not just their ability to correctly respond to a mathematics question. So, for example, a single mathematical problem may be solved differently by students at different (separable) levels of thinking in a learning trajectory. Further, the instructional ramifications of the earlier theories were often based on transmission views in which facts and skills were presented and then passively absorbed. In comparison, learning trajectories have an interactionalist view of pedagogy.

To further elaborate these differences, we consider the three components of learning trajectories. First, the explication of the goal is important and discernible from previous theories of learning that tended to either (1) apply the same theories and procedures to all domains, ignoring subject matter, or (2) accept the goal as arbitrary, or given by existing standards or curriculum. In contrast, as stated, the goals of our learning trajectories are based on both the expertise of mathematicians and research on students' thinking about and learning of mathematics. Thus, in contrast to most earlier approaches, domain-specific expertise *and* research on students' thinking and learning in that domain play a fundamental role in determining the mathematical goal—the first component of learning trajectories.

Second, the developmental progressions of learning trajectories are much more than linear sequences based on accretion of numerous facts and skills. They are based on a progression of levels of thinking that (as does the goal) reflects the cognitive science view of knowledge as interconnected webs of concepts and skills. Each level is characterized by specific mental objects (concepts) and actions (processes) (e.g., Clements, Wilson, & Sarama, 2004; Steffe & Cobb, 1988). These actions-on-objects are students' main way of operating on, knowing, and learning about the world, including the particular "worlds" of mathematical topics. Specification of these actions-on-objects allows a degree of precision not achieved by previous theoretical and empirical works. Further, the research methods that generate and test these mental models are distinct from methods used in earlier research. Strategies such as clinical interviews are used to examine students' knowledge of the content domain, including conceptions, strategies, intuitive ideas, and informal strategies used to solve problems. The researchers set up a situation or task to elicit pertinent concepts and processes. Once an initial model has been developed, it is tested and extended with teaching experiments, which present specific tasks to individual students with the goal of building models of students' thinking and learning—that is, transitions between levels are the crux of these studies (Steffe, Thompson, & von Glasersfeld, 2000)—another way the articulation of learning trajectories differs from many earlier research programs. Once several iterations of such work indicate substantive stability, it is accepted as a working model. Thus, the developmental progressions' levels of thinking and explication of transitions between models describe in detail the following: (1) what students are able to do, (2) what they are not yet able to do but should be able to learn, and (3) why—that is, how they think at each level and how they achieved these levels of thinking. This sets learning trajectories' developmental progressions apart from earlier efforts to develop educational sequences, that, for example, often used reductionist techniques to break a goal regarding competence into subskills, based on an adult's perspective.

Third, the instructional tasks of learning trajectories are much more than didactic presentations or external models of the mathematics to be learned. They often include these elements, but they are fine-tuned to engender the specific actions-on-objects that constitute the level of thinking ($n + 1$) that a particular student (at level n) needs. Learning trajectories differ from instructional designs based on task (or rational) analysis because they are not a reduction of the skills of experts but are models of students' learning that include the unique constructions of students and require continuous, detailed, and simultaneous analyses of goals, pedagogical tasks, teaching, and students' thinking and learning. Such explication allows the researcher to test the theory by testing the curriculum (Clements & Battista, 2000), usually with teaching and design experiments. These scientific experiments include conceptual analyses and theories that "do real design work in generating, selecting and validating design alternatives at the level at which they are consequential for learning" (diSessa & Cobb, 2004, p. 77). They also are tested and iteratively revised in progressively expanding social situations, which results in greater contributions to both educational theory and practice (Clements, 2007).

HOW VALID ARE TODAY'S LEARNING TRAJECTORIES? THE CASE OF EARLY MATHEMATICS

We initially reviewed research in early mathematics because we believed that learning trajectories should be the backbone of our *Building Blocks* research-and-development curriculum project (Clements & Sarama, 1998), which was developed based on a curriculum research framework (Clements, 2007) that puts learning trajectories at the core. Our work in that and several subsequent projects convinced us of the usefulness of the construct. Here, we briefly describe our initial developmental and empirical work and then make several observations based on a recent comprehensive research review of learning trajectories in early mathematics.

Empirical Support for the Building Blocks and TRIAD Instantiations

Building Blocks, funded by the National Science Foundation, was a project that researched and developed software-enhanced pre-kindergarten (pre-K) to grade 2 mathematics curricula. Based on theory and research on early childhood learning and teaching (Bowman, Donovan, & Burns, 2001; Clements, 2001), we determined that the basic approach of *Building Blocks* would be finding the mathematics in and developing mathematics from children's activity. To do so, all aspects of the *Building Blocks* project are based on learning trajectories.

Building Blocks Pre-K consists of a comprehensive curriculum, a teachers' edition, assesment and resource guides, manipulatives, big books, and software. Each component is structured around learning trajectories. A series of studies have supported its effectiveness.

One experiment involved all children in four experimental and comparison classrooms, including state-funded and Head Start Pre-K programs serving low-income children (Clements & Sarama, 2007c). Children in all classrooms were pre- and posttested individually based on the curriculum's hypothesized developmental progressions. The experimental treatment group score increased significantly more than the comparison group score. Effect sizes comparing posttest scores of the experiment group to the comparison group were .85 for number and 1.44 for geometry (Cohen's *d*). Thus, achievement gains of the experimental group were comparable to the sought-after 2-sigma effect of individual tutoring (Bloom, 1984). In summary, this curriculum was shown to be effective under controlled conditions.

Two additional studies took *Building Blocks Pre-K* to scale using the TRIAD model (Sarama, Clements, Starkey, Klein, & Wakeley, 2008). The first study involved 36 teachers randomly assigned to one of three conditions. The first condition was *Building Blocks Pre-K.* The second, comparison condition involved another pre-K math curriculum, with teaching receiving equivalent curriculum resources and equal time in professional development, but without an emphasis on learning trajectories in either the curriculum or the corresponding professional development. The third condition was a business-as-usual control (Clements & Sarama, 2008). The *Building Blocks* group's score increased from pretest to posttest significantly more than the comparison (effect size of .47) and the control group scores (effect size of 1.07). Thus, effect sizes were only slightly lower without the controlled conditions, or "hyperrealization," of the previously described study (i.e., .85 for number and 1.44 for geometry relative to the control group, compared to 1.07 for all topics in this study).

The second, more recent large-scale study, a cluster randomized trial, involved 106 teachers and 1,305 children in two states (Sarama & Clements, 2009b). The effect size of the *Building Blocks Pre-K* curriculum was .72. Note that an external review of the research on *Building Blocks Pre-K* has been conducted with a favorable conclusion by the U.S. Department of Education's "What Works Clearinghouse," using its rigorous criteria (http://ies.ed.gov/ncee/wwc/reports/early_ed/sra_prek/index.asp).

Review of the Literature—General Issues

These empirical studies have provided support for the efficacy and efficiency of basing curricula and professional development on learning trajectories. A limitation of these studies is that each one involved only one curriculum based on learning trajectories. Both to contribute to the research literature and to answer those who wished to use the learning trajectories but did not want to purchase a curriculum, we conducted a research review to describe how reliable and broadly known empirical results can be used to create learning trajectories for mathematics education for an age range of birth to grade 2.

This comprehensive research review was published as a set of two companion books. The first (Sarama & Clements, 2009a) reviews the research underlying our learning trajectories, and especially the developmental progressions, in detail. The second (Clements & Sarama, 2009) describes and illustrates how these learning trajectories can be implemented in the classroom. Although this research review cannot provide the validation of teachers' and students' learning on par with the extensive evidence from the series of *Building Blocks* studies, it comprehensively surveyed the field. Here, we reflect on what the review revealed about the current state of research-based learning trajectories.

The review revealed considerable support for the use of learning trajectories in mathematics education and mixed, but mostly promising bodies of research on which to build learning trajectories for specific topics. We will address each in turn.

The Use of Learning Trajectories

Usefulness to teachers

The pedagogical value of hypothetical learning trajectories has been supported by research (Bredekamp, 2004; Clements & Sarama, 2004a; Simon, 1995). Knowledge of developmental progressions enables high-quality teaching based on understanding both mathematics and students' thinking and learning. Early childhood teachers' knowledge of such mathematical development is related to their students' achievement (Carpenter, Fennema, Peterson, & Carey, 1988; Peterson, Carpenter, & Fennema, 1989). In one study, the few teachers who actually led in-depth discussions in mathematics classrooms saw themselves not as moving through a curriculum, but as helping students move through levels of understanding (Fuson, Carroll, & Drueck, 2000).

Professional development

Multiple researchers have found that professional development focused on developmental progressions results in increases not only in teachers' professional knowledge but also in their students' motivation and achievement (B. A. Clarke, 2004; D. M. Clarke et al., 2001; D. M. Clarke et al., 2002; Fennema, Carpenter, Frank, Levi, Jacobs, & Empson, 1996; Kühne, van den Heuvel-Panhulzen, & Ensor, 2005; Thomas & Ward, 2001; Wright, Martland, Stafford, & Stanger, 2002). Learning trajectories, therefore, can facilitate developmentally appropriate teaching and learning for all students (c.f. Brown, Blondel, Simon, & Black, 1995).

Instructional tasks

The correlated instructional tasks are also important. A description of students' development is essential but is insufficient by itself: It can inform teachers as to what types of thinking to look for and facilitate, but not how to facilitate students' development of the next level. The entire learning trajectory explains the levels of thinking, the mental ideas and actions that must be built, the processes that will engender those ideas and actions (e.g., promote learning), and specific instructional tasks and teaching strategies based on those processes. Tasks are often designed to present a problem that is just beyond the students' present level of operating, so they must actively engage in reformulating the problem or their solution strategies, often with peers and teacher guidance. In reflecting on their activity, they see whether they have solved the original problem or need to engage in more thinking. This cycle may continue until a new level of thinking is built. Finally, a focus on both big ideas and the conceptual storylines of curricula in the form of hypothetical learning trajectories, such as those from the *Building Blocks* project, is supported by research on systemic reform initiatives (Heck, Weiss, Boyd, & Howard, 2002).

Issues and Questions

Before we turn to specific learning trajectories, we address several issues and questions that have emerged in our discussions and presentations (the following is adapted from Clements & Sarama, 2009). Some of them can be answered briefly, but some touch on theoretical and practical issues profoundly, and thus the answers are extended.

When are students "at" a level? Based on the available theoretical and empirical research corpus, we hypothesize that students are at a certain level when most of their behaviors reflect the thinking, concepts, and skills of that level. Often, they show behaviors from the next (and previous) levels as they learn. Recall the illustration of this in Figure 1.1. This takes us to the next question.

Can students work at more than one level at the same time? Although most students work mainly at one level or are in transition between two levels (if they are tired or distracted, they may even operate at a much lower level than they might otherwise), levels are not absolute stages. They are benchmarks of complex growth that represent distinct ways of thinking. So, another way to think of them is as a sequence of different patterns of thinking and reasoning (cf. Cobb, Confrey, diSessa, Lehrer, & Schauble, 2003). Students are continually learning within levels and making transitions to higher levels. They do work mainly at one level and work on achieving the next.

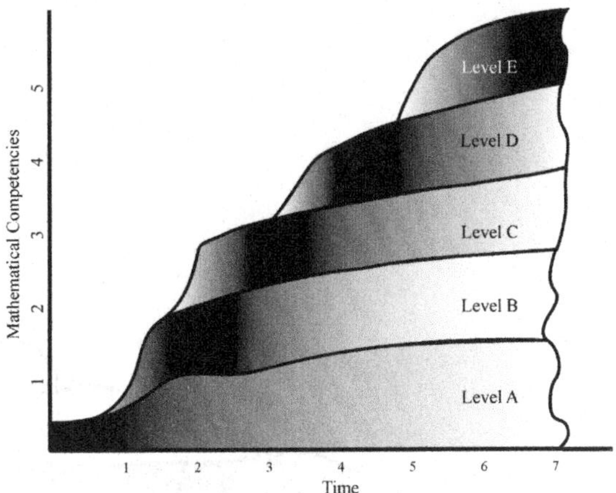

FIGURE 1.2. Developmental progression with good instruction
Note: Hypothetical, idealized view of a student's progression through a developmental progression with good instruction (a valid learning trajectory, with other educational conditions ideal) organized into "units" spread through several elementary grades. The units for both axes are arbitrary and provided only to facilitate comparisons for Figures 1.2-1.4. For this example, time might be considered 6-month units; however, for a specific learning trajectory goal, the unit might be a week.

How do educative experiences affect a student's actual developmental progression? Again, distinct from Piagetian stages, students' progress is dependent on educative experiences (Dewey, 1938/1997). The periods of fast growth (large slopes) illustrated in Figure 1.2 represent a continuous instructional unit focusing on that particular learning trajectory (with patterns of thinking developed at an increasing pace at higher levels, and little forgetting between levels, due in part to idealized integrated review during other units, and few noninstructional interruptions).

In contrast, Figure 1.3 represents the notion that negative experiences can be miseducative (Dewey, 1938/1997). If instruction is confusing or emphasizes routine procedural skills while deemphasizing conceptual learning, including reduction of level (van Hiele, 1986), students' patterns of thinking develop differently. First, there is less learning. Second, this learning can degrade over time. Third, higher levels may be learned partially, but not completely. Fourth, higher levels may not achieve dominance.

Can students skip a level? This is possible in several ways. One way students may appear to skip levels is if there are separate subtrajectories, or subtopics, within a larger learning trajectory. For example, we combined many counting competencies into one counting sequence with subtrajectories,

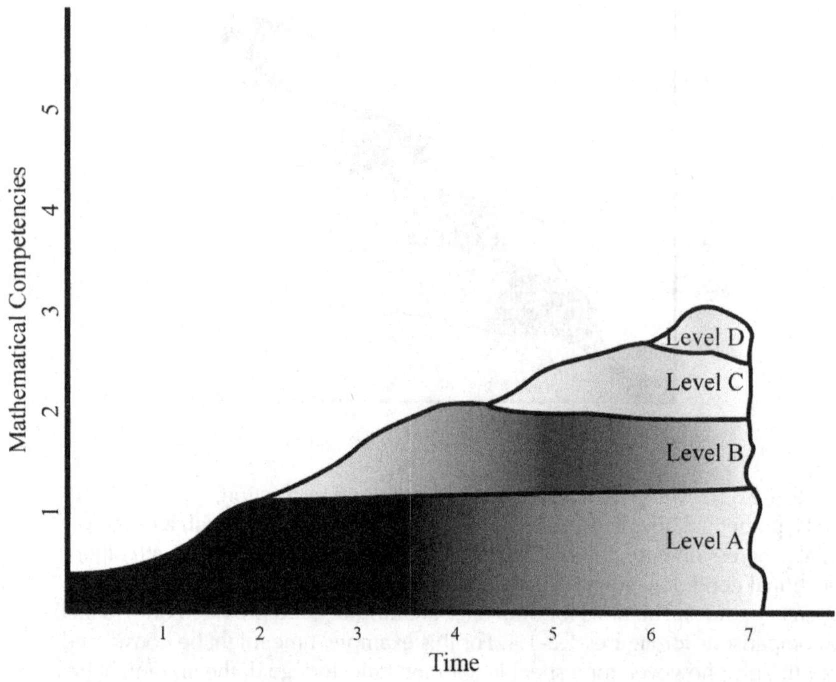

FIGURE 1.3. Developmental progression with inadequate instruction
Note: Hypothetical view of a student's progression through a developmental progression with inadequate instruction throughout the elementary grades.

such as verbal counting skills. Most students learn to count to 100 at age 6 after learning to count objects to 10 or more, so we ordered the learning trajectory in that way. However, some students may learn this verbal skill earlier. The subtrajectory of verbal counting skills would still be followed, and there is no theoretical inconsistency in the superordinate learning trajectory for counting if they do so.

More important, students may appear to skip a level in some cases; for example, they may apparently move from level n to level $n + 2$. As stated previously, students may have been building and operating on more than one level. Then, in a short period of rapid learning or insight, they may progress to a substantially higher level.

This raises the important theoretical issue of the nature of levels of thinking in learning trajectories. There is nothing in the theoretical construct of a learning trajectory in hierarchic interactionalism that prohibits such a developmental pattern. Again, levels are not stages. *Stages,* in Piagetian theory (Piaget & Inhelder, 1967; Piaget & Szeminska, 1952) are long periods of development, characterized by cognition across a variety of domains, that are qualitatively different from that of both the preceding and succeeding stages (Karmiloff-Smith, 1984). In Piagetian theory, stage $n + 2$ requires passage through stage $n + 1$ because stage $n + 1$ constructed the elements from which stage $n + 2$ would be built. (See examples from Barrett, Clements, and Sarama's work in Chapter 4, this volume, for theoretically consistent illustrations of different students' movement through a particular learning trajectory for linear measurement.)

A *level* in a learning trajectory represents a period of qualitatively distinct cognition, as do stages; however, there are at least four important distinctions between levels and stages.

1. First and most important, levels apply only *within* a specific domain rather than across domains.
2. The period of time is generally far shorter, and can be months or even days (especially given efficacious instruction), rather than stages' period of years.
3. While hierarchic interactionalism postulates that subsequent levels are built upon earlier level, the order of magnitude of difference in durations indicates a distinctly different cognitive "distance" between successive states. Informally, the transition between contiguous levels is far smaller than the jump between Piagetian stages. (Admittedly, measuring such distances, for this distinction and for related theoretical notions such as Vygotsky's Zone of Proximal Development, remains an open problem.) Thus, skipping or "jumping over" a level—whether due, for instance, to students; construction of actions-on-objects of two levels simultaneously or lack of opportunity on the part of teachers or researchers to observe a student's construction of level $n + 1$ (so that observations capture only levels n and $n + 2$)—is theoretically acceptable and probable.
4. In a similar vein, the hierarchic interactionalism theory of levels makes no commitment (as does the Piagetian theory of stages) that the actions-on-objects of level $n + 2$ must be built from those of level $n + 1$. Rather, in many cases the cognitive material may be present at level n, requiring only a greater degree of construction or generalization to construct the pattern of thinking and reasoning defining level $n + 2$. Figure 1.4 illustrates a student's construction of multiple levels of thinking over a very curtailed time frame within a rich unit of instruction.

The steep slopes of levels 3, 4, and 5 in Figure 1.4 indicate rapid progression of an individual student through those levels. These suggest another way a student may appear to skip a level—in such periods of quick learning, it may be difficult or impossible for an instructor or teacher to perceive or

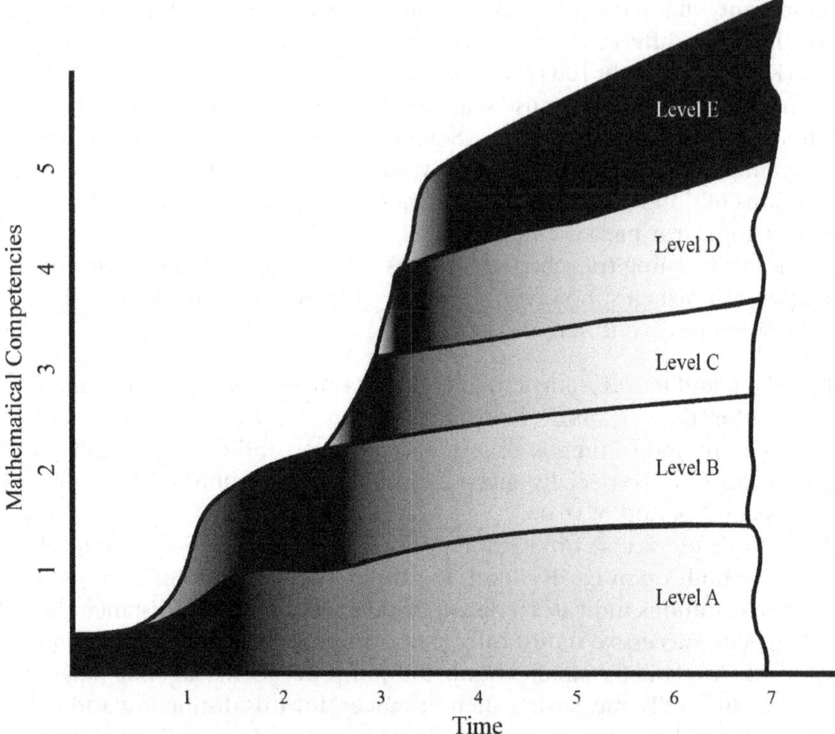

FIGURE 1.4. Developmental progression with ideal conditions
Note: Hypothetical view of student's rapid progress through a developmental progression with ideal conditions: high-quality informal but intentional experiences in the early years and an extended period (large slope) of high-quality, focused instruction with a well-motivated and prepared student population.

document the level(s) at which the student is operating or transitions between levels. That is, two consecutive observations suggesting that a student skipped from level n to level $n + 2$ may constitute an accurate description of the student's learning or simply a lack of observations (or opportunities to observe) of level $n + 1$ thinking that occurred in between the available observations.

Do students ever "fall back" to earlier levels? What does this mean? Although levels of thinking can be theoretically viewed as nonrecurrent (Karmiloff-Smith, 1984), students not only can, but frequently do, return to earlier levels of thinking in certain contexts. Therefore, hierarchic interactionalism postulates the construction of nongenetic levels (Clements et al., 2001), which has two special characteristics:

1. Progress through nongenetic levels is determined more by social influences and, specifically, instruction, than by age-linked development. (At this point, this implies only that progression does not merely occur as a simplistic function of time or biological maturation, but requires instructional intervention. It must be noted however that this does not rule out certain levels developing under maturational constraints.)
2. Although each higher nongenetic level builds on the knowledge that constitutes lower levels, its nongenetic nature means that the instantiation and application of earlier levels in certain contexts is not precluded. Such contexts are often, but are not limited to, especially demanding or stressful contexts or tasks. Earlier levels might also be instantiated in informal settings, such as a determining that a window "is a rectangle" simply because it fits a visual prototype (lower-level, "visual" thinking, as in van Hiele, 1986, but appropriate in the situation).

There exists a probability of evoking each level depending on circumstances. Again, Figure 1.1 illustrates that earlier levels do not disappear—people do not jump from one type of thinking to a separate type, but rather build new ways of thinking upon the previous patterns of thinking. This process is codetermined by conscious metacognitive control, which should increase as each student proceeds up through the levels. Therefore, students have increasing choice to override the default probabilities. The use of different levels is environmentally adaptive: The adjective *higher* should be understood as a higher level of abstraction and generality in students' competencies, without necessarily the implication of either inherent superiority or the abandonment of lower levels as a consequence of the development of higher levels of thinking. Nevertheless, the levels constitute qualitative changes in behavior for those tasks requiring greater abstraction and generality, especially in regard to the construction of mathematical schemes out of action.

Are the instructional tasks in a learning trajectory the only way to teach students to achieve higher levels of thinking? No, there are many ways. In some cases, however, there is some research evidence that these are especially effective ways. Even then, they are just a single best-case scenario. In other cases, they are simply illustrations of the kind of activity that would be appropriate to reach that level of thinking. Further, teachers need to use a variety of pedagogical strategies in teaching the content, presenting the tasks, guiding students in completing them, and so forth.

Are learning trajectories strictly or solely linear? Or would a more complex description be more appropriate? We believe the answer is "both." Learning trajectories capture much about learning within a single topic, especially because much of that learning is like traveling along a path—when that path conveys movement from one complex level of thinking, including interwoven concepts and skills, to the next. This is not simple ladder-like incremental accretion. A better simile would be that learning is like moving from one land or city to another. Further, multiple paths are often possible and predicted by hierarchical interactionalism (Clements et al., 2001; Clements & Sarama, 2007b; Sarama & Clements, 2009a).

However, learning mathematics also involves interactions between topical content knowledge and general mathematics processes (e.g., making connections, problem solving), between topics,[2] between mathematics and other subject areas, and so forth. General educational goals, including "habits of mind" such as imagination, inventiveness, risk taking, and persistence, are also important components. Complete illustrations of students' wonder, excitement, and especially thinking and problem solving (e.g., Clements & Sarama, 2009) are necessary antidotes to the possible misconception that the main story of learning trajectories is students marching through a series of learning levels. Such a dreary picture is the opposite of what learning trajectories are designed to engender, namely, students filled with curiosity and creative ideas and teachers excited about helping them see the world through mathematical lenses.

Also, those who emphasize the importance of big ideas and interconnections of these ideas (Baroody et al., 2004) have an important point. Here we emphasize that important connections should also be established among the goals of various learning trajectories. These are often critical, high-level connections between such big ideas as *decomposition* or *unit,* and should be emphasized in the superordinate curriculum of which learning trajectories are components.

Further, we have found that learning trajectories are constructed at a grain size and level of complexity that is advantageous for teachers. More structurally complex models are easily generated, but they tend to overwhelm teachers. For example, three-dimensional (or higher-dimensional) models are usually as intractable and therefore as useless as the cliché "everything connects to everything." In contrast, learning and thinking about one or two learning trajectories, and their connections to each other *and* to other domains as described above, has proven effective in our own research and that of others.

Are learning trajectories at all age/grade levels similar, and of similar difficulty to construct? Our research review pertained to the early childhood age range. We believe that this age range is at present the most accommodating to the construction of learning trajectories, for at least four reasons.

1. Young children's learning of mathematics has attracted not only researchers from the mathematics education community, but—be-

cause these are the years of initial development of well-defined, foundational ideas (Piaget, 1964; Piaget & Szeminska, 1952)—researchers from the fields of genetic epistemology, developmental and cognitive psychology, and many others. Therefore, there is a rich literature on which to draw to support the construction of developmental sequences.

2. Arguably, the younger the student, the more their learning is influenced by generalizable structures of intuitive knowledge and patterns of thinking and learning, which can thus provide support for the identification of the developmental progressions.
3. These early developing (and probably innate) structures simultaneously support early mathematical development and constrain the paths of that development. In comparison, later mathematics relies at best on secondary, not primary intuitions (Fischbein, 1987). Secondary intuitions and competencies with more sophisticated mathematics must increasingly be educationally engineered.
4. Similarly, the relative simplicity of the concepts and skills at the early childhood level assists the formation of sequences and interrelationships between topics. In contrast, the more advanced the mathematics, the more difficult such formation is, due to the increased complexity of the topics and the geometric progression of the number of ways topics and subtopics might be related and sequenced.

In conclusion, we believe that learning trajectories can and should be constructed to guide standards, curricula, and pedagogy at every level, and they have proved successful across the grades (e.g., other chapters in this volume, as well as Asiala, Dubinsky, Mathews, Morics, & Oktac, 1997; Cobb et al., 2003; Gravemeijer, 1999; Leron & Dubinsky, 1995). It is unsurprising, however, that this construct is less difficult and has proceeded further at the early childhood age range than for older students.

Review of the Literature—Specific Learning Trajectories in Early Mathematics Education

A final question moves us to the issue of specific learning trajectories, or topics, in early mathematics education: *Are developmental progressions in various topics similar in nature and in their empirical support?* The levels in most of our developmental progressions[3] are levels of thinking—a distinct period of time of qualitatively distinct ways of thinking, as described previously. However, some are merely levels of attainment, similar to a mark on a wall to show a student's height. These signify simply that a student has gained more knowledge. For example, students must learn to name or write more numerals, but knowing more does not require deeper or more complex thinking. In summary, some trajectories are more tightly constrained by natural cognitive development than others, and these constraints are often reinforced by the nature of the mathematical development in a particular domain. That is, mathematics is a highly sequential, hierarchical domain in which certain ideas and skills must be learned before others.

Further, our developmental progressions ranged from those with considerable supportive evidence to descriptions that are a "best professional judgment" of such a progression. Indeed, progressions may be determined by natural ways of learning more in some topics (early number and arithmetic) than others (geometric and spatial competencies). Even at best, developmental progressions are general descriptions and can be affected by and modified in relation to cultural and individual differences.

Analyzing all our learning trajectories in detail is beyond the scope of this chapter. (Such details are available in our reviews; see Clements & Sarama, 2009; Sarama & Clements, 2009a.) We will briefly evaluate a few for the purpose of illustration. For example, we view our learning trajectory for counting as including levels of thinking, whose developmental progressions are constrained by both cognitive development and to a lesser extent by mathematics subject matter, and for which both the developmental progressions and instructional tasks are fairly well documented by empirical research. The learning trajectory for beginning addition and subtraction is similar, but arguably more extensively documented by researchers.

In contrast, the learning trajectory for geometric shapes is based on a developmental progression that is somewhat constrained by both cognitive development and mathematics subject matter, but which may have a sequence that is, more so than counting and arithmetic, determined by culturally based experiences. Similarly, there is a less substantial corpus of research with which to document developmental progressions and instructional tasks in this domain. Shape composition may be guided as much by cognitive development as are the trajectories in the number domain; however, the studies that engender and validate that developmental progression—as targeted as they are—are few in number and from a more limited pool of researchers. The developmental progression was born in observations of children's explorations (Sarama, Clements, & Vukelic, 1996) and refined through a series of clinical interviews and focused observations and quantitative evaluations (Clements, Wilson, & Sarama, 2004). From a lack of competence in composing geometric shapes (children would often not even place shapes touching), children gain abilities to combine shapes—initially through trial and error, and gradually by attributes—into pictures, and finally intentionally synthesize combinations of shapes into new shapes (composite shapes, as when a child purposefully finds two isosceles trapezoids so as to compse them into a regular hexagon). A main instructional task requires children to solve outline puzzles with shapes off and on the computer, a motivating activity (Sales, 1994; Sarama et al., 1996). The objects are shapes and composite shapes, and the actions include creating, duplicating, positioning (with geometric motions), combining, and decomposing both individual shapes (units) and composite shapes (units of units). The characteristics of the tasks require actions on these objects corresponding to each level in the learning trajectory. (For full descriptions with illustrations, see Clements & Sarama, 2009; and Sarama & Clements, 2009.)

FINAL WORDS

In this description, we have taken the perspective of the researcher or the researcher/curriculum developer. We conclude with a reminder that learning trajectories are hypothetical and should be reconceptualized and recreated by small groups or individual teachers, so that their actual instantiation is based on more intimate knowledge of the particular students involved—their extant knowledge, learning preferences, and engagement in certain task types or contexts. Indeed, a priori learning trajectories are always hypothetical in that the actual learning and teaching, and the teacher's recognition of these, cannot be completely known in advance. The teacher must construct new models of students' mathematics as they interact with students around the instructional tasks, and thus alter their own knowledge of students and future instructional strategies and paths. Thus, the realized learning trajectory is emergent.

These observations, however, constitute a major rationale for the use of learning trajectories, especially in professional development, *not* a disclaimer regarding limited value. Based on their interpretations, teachers conjecture what the student might be able to learn or abstract from the experiences. Similarly, when they interact with the student, teachers also consider their own actions from the student's point of view. We believe that learning trajectories are the most powerful tool teachers can use to do this well. Thus, the benefit for the teacher is to have a well-formed and specific set of expectations about students' ways of learning—a likely path that incorporates the big, worthwhile ideas.

On the research side, a complete hypothetical learning trajectory includes all three aspects—the learning goal, developmental progressions of thinking and learning, and sequence of instructional tasks. The synergism between the latter two aspects has already been described. Less obvious is that their integration can produce novel results, even within the local theoretical fields of psychology and pedagogy. The enactment of an effective, complete learning trajectory may actually alter developmental progressions or expectations previously established by psychological studies, because they open up new paths for learning and development. This, of course, reflects the traditional, if oversimplified debate between Vygotsky and Piaget regarding the priority of development over learning. We believe that learning trajectory research, along with other research corpora, supports the Vygotskian position that, at least in some domains and some ways, learning and teaching tasks can change the course of development. Such an enactment, based on the fine-grained cognitive analysis of the developmental progression and on the similarly detailed analysis of the instructional tasks, provides a more elaborated theoretical base for curriculum and instruction than is often available, and may also open instructional avenues not previously considered.

Much work remains to be done on the learning trajectories construct. Much will be learned.

AUTHOR NOTE

We appreciate the comments of Jeffrey Barrett on an early draft of this manuscript. This material is based in part upon work supported by the Institute of Education Sciences, U.S. Department of Education, through Grant R305K05157 to the University at Buffalo, State University of New York, D. H. Clements, J. Sarama, and J. Lee, *Scaling Up TRIAD: Teaching Early Mathematics for Understanding with Trajectories and Technologies*, and by the National Science Foundation Research Grant ESI-9730804, *Building Blocks—Foundations for Mathematical Thinking, Pre-Kindergarten to Grade 2: Research-based Materials Development.* Any opinions, findings, and conclusions or recommendations expressed in this publication are those of the authors and do not necessarily reflect the views of the U.S. Department of Education or of the National Science Foundation. It should be noted that, although the research is concerned with the scale-up model (not particular curricula), the curricula evaluated in this research have been published by the authors, who thus could have a vested interest in the results.

NOTES

1. That is, logically defined within the subject matter's discipline, sequences are established by determining what subordinate competences are required by each superordinate competence.
2. Our original and still-preferred representations of learning trajectories were in large tables that showed relationships among all trajectories—tables complex enough to be simplified by most editors! Further, the topical specificity of these learning trajectories actually allows more flexibility than standard scope and sequences. For example, they allow—and show increasing connections between—the chronologically parallel development of counting and arithmetic competencies.

3. Recall that the learning trajectories we describe in the research reviews (Clements & Sarama, 2009; Sarama & Clements, 2009b) were developed as part of the *Building Blocks* project. Many researchers and developers are developing their own.

REFERENCES

Asiala, M., Dubinsky, E., Mathews, D. M., Morics, S., & Oktac, A. (1997). Development of students' understanding of cosets, normality and quotient groups. *Journal of Mathematical Behavior, 16*(3), 241–309.

Baroody, A. J. (1987). *Children's mathematical thinking.* New York, NY: Teachers College.

Baroody, A. J., Cibulskis, M., Lai, M.-l., & Li, X. (2004). Comments on the use of learning trajectories in curriculum development and research. *Mathematical Thinking and Learning, 6,* 227–260.

Baroody, A. J., Lai, M.-L., & Mix, K. S. (2005, December). *Changing views of young children's numerical and arithmetic competencies.* Paper presented at the National Association for the Education of Young Children, Washington, DC.

Beilin, H. (1984). Cognitive theory and mathematical cognition: Geometry and space. In B. Gholson & T. L. Rosenthanl (Eds.), *Applications of cognitive-developmental theory* (pp. 49–93). New York, NY: Academic Press.

Bloom, B. S. (1984). The 2-sigma problem: The search for methods of group instruction as effective as one-to-one tutoring. *Educational Researcher, 13,* 4–16.

Bowman, B. T., Donovan, M. S., & Burns, M. S. (Eds.). (2001). *Eager to learn: Educating our preschoolers.* Washington, DC: National Academy Press.

Bredekamp, S. (2004). Standards for preschool and kindergarten mathematics education. In D. H. Clements, J. Sarama, & A.-M. DiBiase (Eds.), *Engaging young children in mathematics: Standards for early childhood mathematics education* (pp. 77–82). Mahwah, NJ: Lawrence Erlbaum Associates.

Brown, M., Blondel, E., Simon, S., & Black, P. (1995). Progression in measuring. *Research Papers in Education, 10*(2), 143–170.

Brownell, W. A. (1928). *The development of children's number ideas in the primary grades.* Chicago, IL: Department of Education, University of Chicago.

Brownell, W. A., & Moser, H. E. (1949). *Meaningful vs. mechanical learning: A study in grade III subtraction.* Durham, NC: Duke University Press.

Carpenter, T. P., Fennema, E. H., Franke, M. L., Levi, L., & Empson, S. B. (1999). *Children's mathematics: Cognitively guided instruction.* Portsmouth, NH: Heinemann.

Carpenter, T. P., Fennema, E. H., Peterson, P. L., & Carey, D. A. (1988). Teachers' pedagogical content knowledge of students' problem solving in elementary arithmetic. *Journal for Research in Mathematics Education, 19,* 385–401.

Carpenter, T. P., Franke, M. L., Jacobs, V. R., Fennema, E. H., & Empson, S. B. (1998). A longitudinal study of invention and understanding in children's multidigit addition and subtraction. *Journal for Research in Mathematics Education, 29,* 3–20.

Carpenter, T. P., & Moser, J. M. (1984). The acquisition of addition and subtraction concepts in grades one through three. *Journal for Research in Mathematics Education, 15,* 179–202.

Case, R., & Okamoto, Y. (1996). The role of central conceptual structures in the development of children's thought. *Monographs of the Society for Research in Child Development, 246*(6), 1–26.

Clarke, B. A. (2004). A shape is not defined by its shape: Developing young children's geometric understanding. *Journal of Australian Research in Early Childhood Education, 11*(2), 110–127.

Clarke, D. M., Cheeseman, J., Clarke, B., Gervasoni, A., Gronn, D., Horne, M., McDonough, A., & Sullivan, P. (2001). Understanding, assessing and developing young children's mathematical thinking: Research as a powerful tool for professional growth. In J. Bobis, B. Perry, & M. Mitchelmore (Eds.), *Numeracy and beyond (Proceedings of the 24th Annual Conference of the Mathematics Education Research Group of Australasia, Vol. 1)* (pp. 9–26). Reston, Australia: MERGA.

Clarke, D. M., Cheeseman, J., Gervasoni, A., Gronn, D., Horne, M., McDonough, A., Montgomery, P., & Rowley, G. (2002). *Early numeracy research project final report.* Melbourne, Australia: Department of Education, Employment and Training, the Catholic Education Office (Melbourne), and the Association of Independent Schools Victoria.

Clements, D. H. (1997). (Mis?)constructing constructivism. *Teaching Children Mathematics, 4*(4), 198–200.

Clements, D. H. (2001). Mathematics in the preschool. *Teaching Children Mathematics, 7,* 270–275.

Clements, D. H. (2007). Curriculum research: Toward a framework for 'research-based curricula.' *Journal for Research in Mathematics Education, 38,* 35–70.

Clements, D. H., & Battista, M. T. (2000). Designing effective software. In A. E. Kelly & R. A. Lesh (Eds.), *Handbook of research design in mathematics and science education* (pp. 761–776). Mahwah, NJ: Lawrence Erlbaum Associates.

Clements, D. H., Battista, M. T., & Sarama, J. (2001). Logo and geometry [Monograph]. *Journal for Research in Mathematics Education, 10.*

Clements, D. H., & Sarama, J. (1998). *Building blocks—Foundations for mathematical thinking, pre-kindergarten to Grade 2: Research-based materials development.* Buffalo, NY: State University of New York at Buffalo. [National Science Foundation, grant number ESI-9730804; see www.gse.buffalo.edu/org/buildingblocks/].

Clements, D. H., & Sarama, J. (Eds.). (2004a). Hypothetical learning trajectories [Special issue]. *Mathematical Thinking and Learning, 6*(2).

Clements, D. H., & Sarama, J. (2004b). Learning trajectories in mathematics education. *Mathematical Thinking and Learning, 6*(2), 81–89.

Clements, D. H., & Sarama, J. (2007a). *Building blocks—SRA real math teacher's edition, grade preK.* Columbus, OH: SRA/McGraw-Hill.

Clements, D. H., & Sarama, J. (2007b). Early childhood mathematics learning. In F. K. Lester, Jr. (Ed.), *Second handbook of research on mathematics teaching and learning* (Vol. 1, pp. 461–555). Charlotte, NC: Information Age Publishing.

Clements, D. H., & Sarama, J. (2007c). Effects of a preschool mathematics curriculum: Summative research on the *Building Blocks* project. *Journal for Research in Mathematics Education, 38,* 136–163.

Clements, D. H., & Sarama, J. (2008). Experimental evaluation of the effects of a research-based preschool mathematics curriculum. *American Educational Research Journal, 45,* 443–494.

Clements, D. H., & Sarama, J. (2009). *Learning and teaching early math: The learning trajectories approach.* New York, NY: Routledge.

Clements, D. H., Sarama, J., & DiBiase, A.-M. (2004). *Engaging young children in mathematics: Standards for early childhood mathematics education.* Mahwah, NJ: Lawrence Erlbaum Associates.

Clements, D. H., Wilson, D. C., & Sarama, J. (2004). Young children's composition of geometric figures: A learning trajectory. *Mathematical Thinking and Learning, 6,* 163–184.

Cobb, P., Confrey, J., diSessa, A., Lehrer, R., & Schauble, L. (2003). Design experiments in educational research. *Educational Researcher, 32*(1), 9–13.

Confrey, J., & Kazak, S. (2006). A thirty-year reflection on constructivism in mathematics education in PME. In A. Gutiérrez & P. Boero (Eds.), *Handbook of research on the psychology of mathematics education: Past, present, and future* (pp. 305–345). Rotterdam, The Netherlands: Sense Publishers.

Davis, R. B. (1984). *Learning mathematics: The cognitive science approach to mathematics education.* Norwood, NJ: Ablex.

Dewey, J. (1997). *Experience and education.* New York, NY: Simon & Schuster. (original work published 1938)

diSessa, A. A., & Cobb, P. (2004). Ontological innovation and the role of theory in design experiments. *Journal of the Learning Sciences, 13*(1), 77–103.

Dowker, A. (2005). Early identification and intervention for students with mathematics difficulties. *Journal of Learning Disabilities, 38,* 324–332.

Fennema, E. H., Carpenter, T. P., Frank, M. L., Levi, L., Jacobs, V. R., & Empson, S. B. (1996). A longitudinal study of learning to use children's thinking in mathematics instruction. *Journal for Research in Mathematics Education, 27,* 403–434.

Fischbein, E. (1987). *Intuition in science and mathematics.* Dordrecht, the Netherlands: D. Reidel.

Fuson, K. C. (2004). Pre-K to grade 2 goals and standards: Achieving 21st century mastery for all. In D. H. Clements, J. Sarama, & A.-M. DiBiase (Eds.), *Engaging young children in mathematics: Standards for early childhood mathematics education* (pp. 105–148). Mahwah, NJ: Lawrence Erlbaum Associates.

Fuson, K. C., Carroll, W. M., & Drueck, J. V. (2000). Achievement results for second and third graders using the standards-based curriculum *Everyday Mathematics. Journal for Research in Mathematics Education, 31,* 277–295.

Gagné, R. M. (1965). *The conditions of learning.* New York, NY: Holt, Rinehart & Winston.

Gagné, R. M., & Briggs, L. J. (1979). *Principles of instructional design.* New York, NY: Holt, Rinehart & Winston.

von Glasersfeld, E.. (1995). Sensory experience, abstraction, and teaching. In L. P. Steffe & J. Gale (Eds.), *Constructivism in education* (pp. 369–383). Mahwah, NJ: Lawrence Erlbaum Associates.

Gravemeijer, K. P. E. (1999). How emergent models may foster the constitution of formal mathematics. *Mathematical Thinking and Learning, 1,* 155–177.

Greeno, J. G., Riley, M. S., & Gelman, R. (1984). Conceptual competence and children's counting. *Cognitive Psychology, 16,* 94–143.

Greenough, W. T., Black, J. E., & Wallace, C. S. (1987). Experience and brain development. *Children Development, 58,* 539–559.

Griffin, S., & Case, R. (1997). Re-thinking the primary school math curriculum: An approach based on cognitive science. *Issues in Education, 3*(1), 1–49.

Heck, D. J., Weiss, I. R., Boyd, S., & Howard, M. (2002). *Lessons learned about planning and implementing statewide systemic initiatives in mathematics and science education.* Retrieved from http://www.horizon-research.com/public.htm

Hershkowitz, R., & Dreyfus, T. (1991). Loci and visual thinking. In F. Furinghetti (Ed.), *Proceedings of the fifteenth annual meeting International Group for the Psychology of Mathematics Education* (Vol. II, pp. 181–188). Genova, Italy: Program Committee.

Hoz, R. (1979). The use of heuristic models in mathematics teaching. *International Journal of Mathematical Education in Science and Technology, 10,* 137–151.

Karmiloff-Smith, A. (1984). Children's problem solving. In M. E. Lamb, A. L. Brown, & B. Rogoff (Eds.), *Advances in developmental psychology* (Vol. 3, pp. 39–90). Mahwah, NJ: Lawrence Erlbaum Associates.

Karmiloff-Smith, A. (1992). *Beyond modularity: A developmental perspective on cognitive science.* Cambridge, MA: MIT Press.

Kilpatrick, J., Swafford, J., & Findell, B. (2001). *Adding it up: Helping children learn mathematics.* Washington, DC: National Academy Press.

Klahr, D., & Wallace, J. G. (1976). *Cognitive development: An information-processing view.* Mahwah, NJ: Lawrence Erlbaum Associates.

Kühne, C., van den Heuvel-Panhulzen, M., & Ensor, P. (2005). From the Lotto game to subtracting two-digit numbers in first-graders. In H. L. Chick & J. L. Vincent (Eds.), *Proceedings of the 29th Conference of the International Group for the Psychology in Mathematics Education* (Vol. 3, pp. 249–256). Melbourne, AU: PME.

Leron, U., & Dubinsky, E. (1995). An abstract algebra story. *American Mathematical Monthly, 102*(3), 227–242.

Lesh, R. A., & Yoon, C. (2004). Evolving communities of mind—in which development involves several interacting and simultaneously development strands. *Mathematical Thinking and Learning, 6*(2), 205–226

Murata, A., & Fuson, K. C. (2006). Teaching as assisting individual constructive paths within an interdependent class learning zone: Japanese first graders learning to add using 10. *Journal for Research in Mathematics Education, 37,* 421–456.

National Council of Teachers of Mathematics. (2006). *Curriculum focal points for prekindergarten through grade 8 mathematics: A quest for coherence.* Reston, VA: Author.

Peterson, P. L., Carpenter, T. P., & Fennema, E. H. (1989). Teachers' knowledge of students' knowledge in mathematics problem solving: Correlational and case analyses. *Journal of Educational Psychology, 81,* 558–569.

Piaget, J. (1964). *The early growth of logic in the child.* New York, NY: W.W. Norton.

Piaget, J., & Inhelder, B. (1967). *The child's conception of space* (F. J. Langdon & J. L. Lunzer, Trans.). New York, NY: W. W. Norton.

Piaget, J., & Szeminska, A. (1952). *The child's conception of number.* London, UK: Routledge and Kegan Paul.

Resnick, L. B. (1994). Situated rationalism: Biological and social preparation for learning. In L. A. Hirschfield & S. A. Gelman (Eds.), *Mapping the mind. Domain-specificity in cognition and culture* (pp. 474–493). Cambridge, MA: Cambridge University Press.

Resnick, L. B., & Ford, W. (1981). *The psychology of mathematics for instruction.* Hillsdale, NJ: Lawrence Erlbaum.

Sales, C. (1994). *A constructivist instructional project on developing geometric problem solving abilities using pattern blocks and tangrams with young children.* Unpublished master's thesis, University of Northern Iowa, Cedar Falls, Iowa

Sarama, J., & Clements, D. H. (2009a). *Early childhood mathematics education research: Learning trajectories for young children.* New York, NY: Routledge.

Sarama, J., & Clements, D. H. (2009b, April). *Scaling up successful interventions: Multidisciplinary perspectives.* Paper presented at the American Educational Research Association, San Diego, CA.

Sarama, J., Clements, D. H., Starkey, P., Klein, A., & Wakeley, A. (2008). Scaling up the implementation of a pre-kindergarten mathematics curriculum: Teaching for understanding with trajectories and technologies. *Journal of Research on Educational Effectiveness, 1,* 89–119.

Sarama, J., Clements, D. H., & Vukelic, E. B. (1996). The role of a computer manipulative in fostering specific psychological/mathematical processes. In E. Jakubowski, D. Watkins & H. Biske (Eds.), *Proceedings of the 18th annual meeting of the North America Chapter of the International Group for the Psychology of Mathematics Education* (Vol. 2, pp. 567–572). Columbus, OH: ERIC Clearinghouse for Science, Mathematics, and Environmental Education

Siegler, R. S., & Alibali, M. W. (2005). *Children's thinking.* Englewood Cliffs, NJ: Prentice-Hall.

Siegler, R. S., & Booth, J. L. (2004). Development of numerical estimation in young children. *Child Development, 75,* 428–444.

Simon, M. A. (1995). Reconstructing mathematics pedagogy from a constructivist perspective. *Journal for Research in Mathematics Education, 26*(2), 114–145.

Steffe, L. P., & Cobb, P. (1988). *Construction of arithmetical meanings and strategies.* New York, NY: Springer-Verlag.

Steffe, L. P., Thompson, P. W., & von Glasersfeld, E.. (2000). Teaching experiment methodology: Underlying principles and essential elements. In A. E. Kelly & R. A. Lesh (Eds.), *Handbook of research design in mathematics and science education* (pp. 267–306). Mahwah, NJ: Lawrence Erlbaum Associates.

Thomas, G., & Ward, J. (2001). *An evaluation of the Count Me In Too pilot project.* Wellington, New Zealand: Ministry of Education.

Thorndike, E. L. (1922). *The psychology of arithmetic.* New York, NY: Macmillan.

Tyler, D., & McKenzie, B. E. (1990). Spatial updating and training effects in the first year of human infancy. *Journal of Experimental Child Psychology, 50,* 445–461.

Van de Rijt, B. A. M., & Van Luit, J. E. H. (1999). Milestones in the development of infant numeracy. *Scandinavian Journal of Psychology, 40,* 65–71.

van Hiele, P. M. (1986). *Structure and insight: A theory of mathematics education.* Orlando, FL: Academic Press.

Wright, R. J., Martland, J., Stafford, A. K., & Stanger, G. (2002). *Teaching number: Advancing children's skills and strategies.* London: Paul Chapman Publications/ Sage.

CHAPTER 2

TOWARD ESTABLISHING A LEARNING PROGRESSION TO SUPPORT THE DEVELOPMENT OF STATISTICAL REASONING

Richard Lehrer, Min-Joung Kim, Elizabeth Ayers, and Mark Wilson

ABSTRACT

Learning progressions articulate prospective means for nurturing the long-term development of disciplinary knowledge and dispositions. Establishing a learning progression is an epistemic enterprise: The goal is to position students to participate in the generation and revision of forms of knowledge valued by a discipline. Here we describe steps taken to establish a learning progression intended to support the development of middle school students' reasoning about data and statistics by engaging them in practices of data modeling. Data modeling refers to the invention and revision of models of chance to describe the variability inherent in particular processes. The construction of the progression involved analysis of core concepts and practices of data modeling that were apt to be generative, yet intelligible, to students; conjectures about fruitful instructional means for supporting student learn-

Learning Over Time: Learning Trajectories in Mathematics Education, pages 31–59.

ing developed in partnership with teachers; and the design and development of an assessment system that could be employed for both summative and formative purposes. Versions of the progression were investigated in a series of classroom design studies, first conducted by the designers of the progression and later by teachers who had not participated in the initial iterations of the design. The movement from initial conjectures to a more stabilized progression involved coordination as well as conflict among disparate professional communities, including teachers, learning researchers, and assessment researchers. Coordination and conflict among communities were mediated by a series of boundary objects, ranging from curriculum units to construct maps.

INITIAL CONSIDERATIONS FOR DESIGNING A LEARNING PROGRESSION

Learning progressions articulate prospective means for nurturing the long-term development of disciplinary knowledge and dispositions, coordinated with conjectured states and transitions in students' developing understandings (National Research Council, 2006). Challenges to establishing a learning progression include the need to parse and distill a discipline to focus on core concepts and practices that are apt to be generative, so that learners are positioned to develop disciplinary dispositions. When we say *disciplinary disposition*, we refer to opportunities for learners to experience what Knorr-Cetina (1999) terms an *epistemic culture*—an arrangement of social, cognitive, and material mechanisms that support disciplinary-distinct ways of knowing (Lehrer, 2009). Hence, at its core, a learning progression is an epistemic enterprise, much in the manner described by Piaget (1970). Learners are involved in the production and revision of knowledge in ways that are consistent with the epistemic practices of the discipline, even if, at interim points along the way, their knowledge does not always conform to readily recognized disciplined forms. This genetic view of knowledge reflects a "commitment that the structures, forms, and possibly the content of knowledge is determined in major respects by its developmental history" (diSessa, 1995, p. 23). Hence, a learning progression cultivates the seeds of disciplinary knowledge and disposition. But it typically also makes distinctions among states of knowledge and forms of practice that may not be acknowledged by the discipline, partly because the discipline has often relegated some of these distinctions to history, effacing them in the process, and partly because learners often make distinctions and must generate coordination among elements of disciplinary knowledge not expressed by the conventions of the discipline.

To cultivate the seeds of this epistemic development implies that learning progressions cannot be constructed from a top-down view of disciplinary progress but must instead be generated as a form of conversation between the discipline and the learner. That is, the progression must characterize states of learners' knowledge even when these states diverge from disciplinary norms yet maintain contact with the concerns and values of the discipline. The dialogue between learner and discipline is mediated by instructional design, which represents a theory about how to coordinate elements of a learning ecology, such as the nature of the problems and/or questions posed to students, to support development in a manner envisioned by the learning progression. Instructional decisions are optimally made in light of evidence about student learning, so they represent conjectures about patterns of typical development of student thinking, mediated by systems of instruction and assessment. Learning, instruction, and assessment circulate as depicted in Figure 2.1. An immediate implication is that a learning progression is not adequately described by any component in isolation.

The image of a learning progression indicated by Figure 2.1 indexes a complex problem of educational design. A progression must be articulated in ways that support the alignment of discipline, learning, instruction, and assessment. An epistemic view of discipline considers how concepts are generated and refined in a field of inquiry. Depictions of learning include descriptions of states of student knowledge, including concepts and practices, and consequential transitions among these states. If learning progressions are to guide teaching, then teachers must be able to identify classes of student performances as representing particular states of the progression. This effort may require innovations in assessment that reveal students' ways

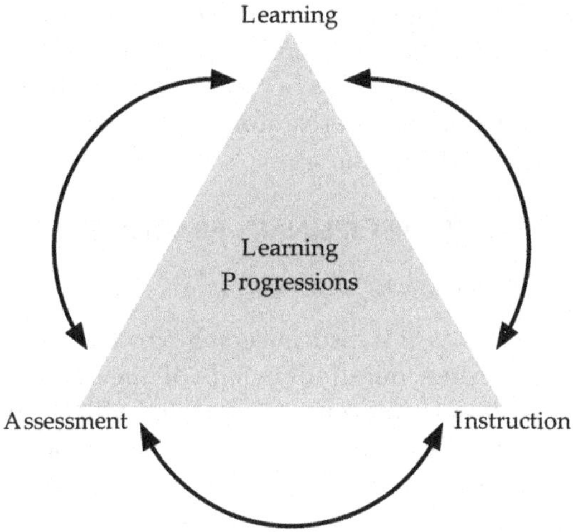

FIGURE 2.1. Learning progressions coordinate learning, instruction, and assessment.

of thinking. And, once student states are identified, teaching practices must come to include a repertoire of appropriate pedagogical responses to support students' progress. The teacher's vision of the disciplinary material to be taught might also need to shift from a logical decomposition of subject matter to a focus on how learners are disposed toward the creation and revision of that subject matter. Hence, rather than conceiving of a learning progression as a map of conceptual change, we find it useful to consider a learning progression as an educational system designed to support particular forms of conceptual change. The components of the system include descriptions of learning, means to support these forms of learning, well-articulated schemes of assessment, and professional development that produces pedagogical capacities oriented toward sustaining student progress.

In this chapter, we describe our steps toward developing a learning progression to support the development of statistical reasoning. The progression is centered at the late elementary and middle school grades but could in principle be extended to encompass younger and older students. It is a work in progress, but one that we are increasingly confident is useful for coordinating learning, instruction, and assessment in this realm. As we describe more completely below, we framed the epistemic practices of data and statistics as a form of modeling, which oriented us toward particular dimensions and states of learner knowledge. These dimensions and states were contested and refined in a series of design studies (Cobb, Confrey, diSessa, Lehrer, & Schauble, 2003), conducted first by researchers and later by teachers who were not part of the original design effort. The design studies were sites for examining relations between means of support—instruction—and learning. To generate evidence about learning, we developed assessments, intending both formative and summative roles for these. Assessment, instruction, and learning became increasingly intertwined as we produced artifacts that mediated relations among professional development, student learning, and assessment.

FRAMING DISCIPLINARY PRACTICES

Data Modeling as a Framework for Instruction

The professional practices of statisticians range from posing questions to making inferences about these questions in light of uncertainty and variability. The path from question to inference is navigated with the development of models—of data, variability, and chance—so that the practice of statistics is a form of modeling, a signature of the sciences (Nersessian, 2008). Schooling, however, especially in the late elementary and middle school grades, typically isolates data from chance and fails to focus on these model-based foundations of statistical reasoning (Cobb, 1999; Garfield, DelMas, & Chance, 2007; Konold & Lehrer, 2008; Moore, 1990; Torok & Watson,

2000; Watson, 2006). To remedy this piecemeal approach toward elements of statistical reasoning, we adopted an alternative perspective, termed *data modeling* (Lehrer & Romberg, 1996), which stresses coordination among these components of statistical reasoning, as illustrated in Figure 2.2. Data modeling begins with inquiry about phenomena and includes the selection of attributes of the phenomena that have the potential to inform the inquiry. Attributes must be defined and measured in light of questions, and the development of ways to measure the attributes often results in revisions to definition of the attributes themselves. Measures set the stage for the generation of data, which must be structured and represented to serve the purposes of inquiry. Statistics measure characteristics of distributed data, and models of chance support inference about these statistics in light of variability inherent in chance events.

The components of Figure 2.2 suggested an initial framing of learning to reason statistically as developing knowledge about each component and as developing relations among the components. For example, understanding a statistic as a measure of a sample distribution is amplified by modeling the sample-to-sample variability of a statistic that results from chance variation. Developing appreciation of random variability helps one understand a statistic as an estimator of a population, rather than as a static description of

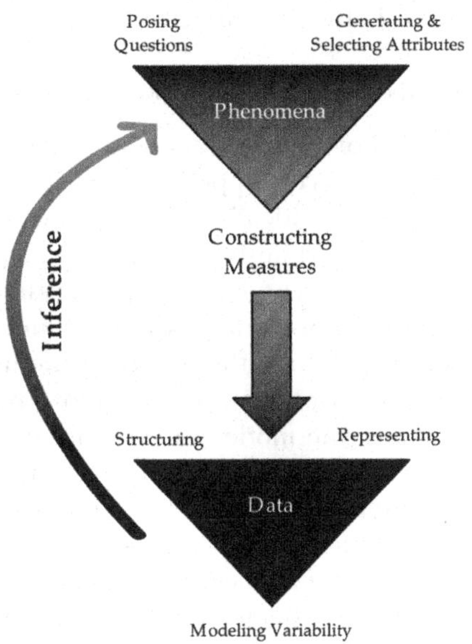

FIGURE 2.2. Data modeling integrating inquiry, data, change, and inference.

a particular sample. Hence, initial conceptions of statistics, perhaps understood as mere computations (Zawojewski & Shaughnessy, 2000), are refined and revised as one engages in practices of measure and model (Lehrer, Kim, & Schauble, 2007).

Relation Between Data Modeling, Teaching, and Learning

Although the framework depicted in Figure 2.2 guided the design of instruction, it failed to specify particular states of knowledge, and it did not indicate prototypical means for supporting the progressive emergence of these states of knowledge. To take these next steps, we first reviewed existing literature about prospective pathways of development and supplemented the comparatively sparse information that emerged from this review by conducting a series of design studies (Cobb et al., 2003) in fifth- and sixth-grade classrooms. One of us served as the primary instructor for these classroom studies. Forms of evidence about relations among teaching, learning, and resulting states of student knowledge ranged from observation of in-the-moment classroom interactions, analysis of video recordings of those interactions, clinical interviews conducted with individual students before and after each cycle of classroom study, and student responses to items that were designed to encapsulate our emerging sense of states of student knowledge. In the next section, we summarize a hypothetical trajectory of student learning that resulted from these multiple iterations and cycles of design (Lehrer et al., 2007; Lehrer & Kim, 2009).

DESIGNING A LEARNING TRAJECTORY

A *learning trajectory* is a series of conjectures about states of student knowledge coordinated with the means of supporting transitions in these knowledge states (Simon, 1995). It is an idealization that is realized in particular settings of instruction and is distinguished from a learning progression by its comparatively reduced scope: The learning progression includes related strands of professional development, assessment, and institutional organization. Means of support include the nature of the tasks or investigations posed; the systems of inscription (representation) employed as tools for conducting investigations; the modes and means by which competing claims are contested and resolved; and participant frameworks, recurrent forms of normative classroom activity, that serve to orchestrate student activity. The orchestration of student activity may be subject to further litmus tests, such as the opportunities that accrue to students to position themselves in relation to the discipline (Boaler, 2002; Gresalfi, 2009) in ways congruent with the epistemic structure of that discipline (Lehrer, 2009). Following multiple iterations of classroom design research, we have settled on a provisional stabilization of a learning trajectory, which we describe next.

Situating Statistical Reasoning in Repeated Measures

Students first generate their own data as each student measures the same attribute of the same object, such as the length of their teacher's outstretched arms, fingertip to fingertip. (We employ measures of arm span for this exposition.) These repeated measures make variability apparent. Wild and Pfannkuch (1999) suggest that "noticing and acknowledging variation" (p. 226) initiates statistical reasoning; the very need for the discipline of statistics arose from the omnipresence of variability (Cobb & Moore, 1997). Student agency is promoted by student participation in the process of measuring, and the repeated measure context helps students coordinate the emergent collection of individual measurements (i.e., data) with ready access to the process that generated the data. The measurement is repeated with another tool, a change in process that affects the variability of the resulting data collection. In sum, in this initial phase, students engage with the upper triangle of Figure 2.2, beginning with a question about the length of their teacher's arm span and generating a collection of data by a process of repeated measure. The first encounter with variability is one of simple difference, because the measures are not identical.

Fostering Representational and Meta-Representational Competencies

Following their generation of the measurements, students begin to structure and represent the collection. Students are challenged, in small groups, to invent paper-and-pencil displays on chart paper that capture their sense of any pattern or trend that they notice in the batch of measured outcomes that their display could communicate to someone else at a glance. The intention is to engage students in the construction of the "shape" of the data. Students generate a variety of displays. Most of the displays bear little resemblance to conventional displays, despite the fact that the students, by this point in their schooling, have usually had several years of exposure to conventional displays. Perhaps this suggests that students are not accustomed to employing displays to address real questions. Different student displays typically highlight different forms of data structure. Classroom conversations are orchestrated to consider how the choices of the inventors result in different appearances or shapes of data, a representational competence (Greeno & Hall, 1997). The forms of representational competence most valued here are those involved in constructing and interpreting displays in ways that reveal their aggregation. To promote meta-representational competence (diSessa, 2004), the teacher guides classroom conversations by asking what each display "shows and hides." That is, how do the choices of the designers highlight some qualities of the data while obscuring others? Teachers do not leave the comparisons of displays to happenstance but instead promote particular classes of comparison, such as comparing a display that partitions continuous data but does not reveal its scale to one that partitions and preserves scale. The latter allows viewers to see holes or gaps in data. Note that comparing displays at a collective level offers opportunities for learning that are not available to individual invention.

Students also consider what might happen if the measurement process were repeated. This image of repeated process supports longer term development of modeling chance (Cobb & Moore, 1997; Moore, 1990; Thompson, Liu, & Saldanha, 2007). Within the shorter term of the lesson, however, the question helps students to coordinate their senses of variability and repetition of a process. Students usually suggest that outliers and other "mistakes" are unlikely to reoccur, because a person or an object has a definitive length. The definitive length is a signal that generates a central tendency in the collection. This is not exactly what is intended by the disciplined view of a repeated, random process, but it is a sufficient seed for developing this image. This is an example of the way in which the seeds of a disciplinary understanding need not mimic mature disciplinary knowledge, because students need not attribute variability to chance at this point in the trajectory.

Statistics Measure Characteristics of Distribution

After considering how displays of data can be constructed and how the choices made by the displays' designers constrain the resulting shape of the data, students are challenged to invent a measure of the "best guess of the real measurement." The purpose is to help students regard statistics as measures of characteristics of a distribution, and it is supported by a conception of measurements as being composed of signal and noise (Konold & Pollatsek, 2002; Petrosino, Lehrer, & Schauble, 2003). Hence, a best guess, a measure of center, has a clear interpretation as the signal in the data—the true length of their teacher's arm span. Students use the displays they generated previously to support their invention of a measure of center, perhaps attempting to somehow capture the center clump of data. After the small groups of students invent measures (statistics), students from other groups attempt to use these invented measures to find the best guess. Instructional conversations are oriented toward identifying which aspects of the batch of data are attended to by particular inventions. The students' invented statistics often have counterparts in disciplinary conventions, such as the mode and the median. By inventing statistics, students have the opportunity to understand relations between statistics and particular qualities of distribution, and more generally, to appreciate the role of algorithm (Berlinski, 2000) in mathematics. Students go on to employ their invention, or one generated by a classmate, to the batch of data generated with the second tool (e.g., a meter stick)—and possibly to other batches of repeated measurement data that have been generated in the classroom. The extension to other sets of data supports the mathematical principle of generalization. Students consider whether the measure yields sensible results in light of what they know about the process that generated the data.

The foregoing process of inventing measure is then repeated, but with the focus specifically on variability. Students invent measures (statistics) of the precision of the batch of measurements. Classroom activity is again directed toward harvesting the fruits of variations in students' inventions and to bring student inventions into contact with conventional statistics of variablity (Lehrer & Kim, 2009; Lehrer, Kim, & Jones, 2011). The representational system, however, is expanded now to include TinkerPlots (Konold & Miller, 2005; Konold, 2007) so that students can capitalize on the affordances of this software tool. These affordances include a suite of tools for structuring and partitioning distributions and for building models of chance. The conceptual systems developed to visualize and measure characteristics of distributions are then extended to other signal-noise processes, such as those involved in production of products (e.g., comparisons of methods for packaging candies).

Modeling Chance

Conceptions of random sampling and the distinction between the distribution of a sample and a sampling distribution are fundamental for inference (Moore, 1990). These forms of reasoning are seeded in the classroom through investigating the variability of the outcomes of simple random devices. It is here that a more disciplinary sense of repeated process is developed, and students are asked, "Despite uncertainty about a single outcome of a repeated process, is there any pattern in the long run?"

Students initially investigate the outcomes of repeated spins of a simple two-region spinner, first generating the outcomes by running the device a particular number of times, and then observing the resulting outcomes and inferring the structure of the device. To make this inference, students redeploy their knowledge of statistics, but now the focus is on sampling variability. Figure 2.3 displays an example investigation. Students see a two-color mystery spinner in TinkerPlots (3a) and then observe the output of 10 repetitions (trials) of the spinner (3b). They decide on a measure of this process, perhaps the percentage of blue outcomes (the statistic), and repeat the sampling process 200 times to develop an *empirical* approximation to the sampling distribution (3c). The central tendency of this statistic estimates the probability of percent-blue. When they uncover the device (3d), students reconcile their empirical estimate with one derived from the proportion of area covered by each region, that is, the *theoretical* estimate of the probability of blue. The behavior of the device helps clarify that both the empirical and the theoretical probabilities are estimates, with the first derived from device behavior and the second derived from analysis of the structure of that device. From these modest beginnings, students investi-

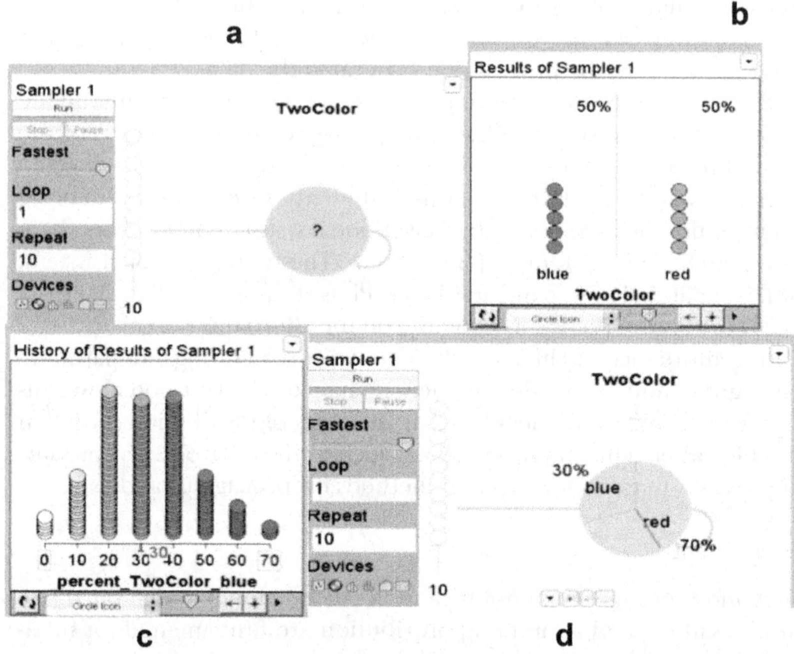

FIGURE 2.3. Simulating sampling distributions to estimate probabilities. Note: Relating theoretical and empirical estimates of probability by simulating sampling distributions.

gate other examples of relations between theoretical and empirical estimates of probabilities of events for both simple events (such as the one described) and compound events (such as the probabilities of the sums of two equally partitioned spinners). These investigations introduce sample spaces as useful tools for estimating probabilities. Sample space estimates of probability, reflecting a theoretical probability structure, are compared to empirical estimates generated by running simulations of devices (and combinations of devices).

Modeling Measurement

Armed with knowledge of the behavior of simple random devices, students analyze potential sources of measurement variability—a student-driven analysis of variance. First, they identify likely sources of error and design a "chance device" for each source. The design of each chance device specifies students' estimates of the magnitudes of error (e.g., + or – 2 cm) and the relative likelihood of each magnitude. Models of the measurement process are represented as a linear combination of the outcomes of the chance

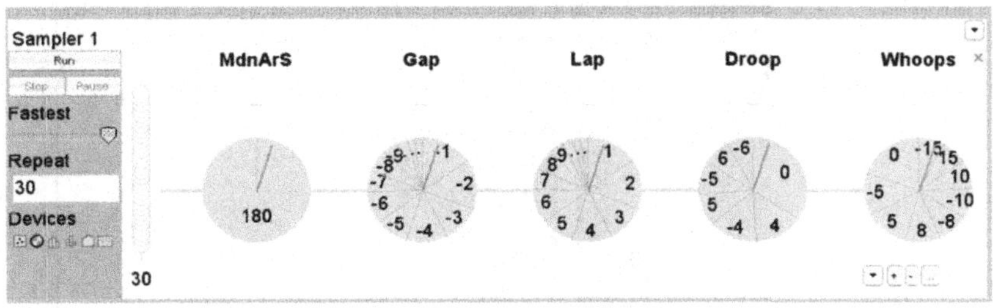

FIGURE 2.4. A student's model of an arm span measure process.

devices and a constant (the best guess) intended to designate the true score (Lehrer et al., 2007). In classical measurement theory, this is the familiar additive model of true score and error. One model of this kind is displayed in Figure 2.4. The first term in the model, *MdnArS,* corresponds to the students' best guess of the arm span of their teacher, the sample median. The remaining spinners depict random sources of measurement error: *Gap* and *Lap* refer to errors of ruler iteration, *Droop* to teacher fatigue (a tendency to drop one's arms under conditions of repeated measure), and *Whoops* to computational mistakes.

As students develop, compare, and contrast their models, they address questions of fit. If the model is run repeatedly to generate a sampling distribution of simulated measurements, does it do a good job? What are good ways of deciding on fit? What is the central tendency of the model? Is it consistent with the observed center of the actual measurements of the teacher's arm span? What is the tendency of the precision of the model measurements? Is this also consistent with the real measurements of the teacher's arm span? Students modify their models to improve them.

Students also create a "bad" model, one with a similar median and range but an implausible shape of the data. Creation of both good and bad models helps students understand the ways that the likelihoods and magnitudes of individual error components can affect the distribution of their composition. The intention in this activity design is to help students see the quality of a particular model in light of other potential models—hence the emphasis on creation of bad models as well as good ones.

Inference

As one might expect from the data modeling perspective, inference is at play in each of the preceding activities: What can be inferred about the data from this display? Is the invented statistic a good measure? What is the likely structure of the mystery spinner? Is the model a good one, considering the data and alternative models? In each case, the purpose of the activity

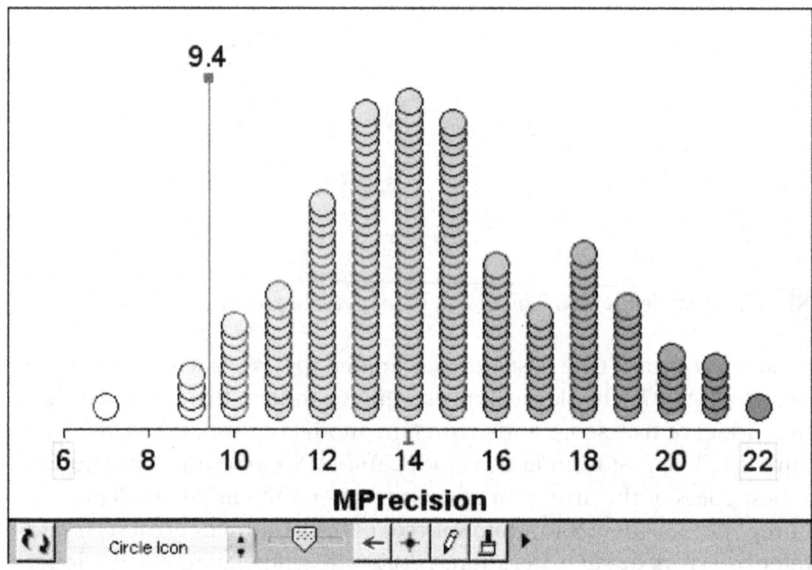

FIGURE 2.5. Empirical sampling distribution for model variability measure. Note: An empirical sampling distribution of a measure of the variability of a model process.

is to make an inference. Our aim is to expose the foundations of statistical inference in modeling and to make its logic apparent, not to replicate the activity of professionals. Hence, in the concluding phase of the instruction, we help students develop a model-based perspective on inference. To do so, we exploit the concepts of models and of sampling distributions of statistics generated by repeated runs of models to pose new questions about the measurements. Students inspect new samples of measurements and make a judgment about whether they were produced by measuring their teacher or, perhaps, by measuring a different teacher. In other scenarios, students evaluate claims about improvement in methods of measure by considering whether or not the precision of measure has indeed increased in light of the sampling variability of this statistic. Figure 2.5 displays the sampling distribution of the precision of a model of the measurements typically obtained using the "standard" method. Then, students decide whether a precision value of 9.4 in a new sample is the result of a change in the method of measurement or, instead, is due just to chance. In professional practice, statistical inference relies on conceptions that students do not yet have, including knowledge of functions that generate distributions consistent with the assumptions made about a process. However, by engaging students in modeling, we invite them into the means by which statistical inferences are produced. This is an example of a state of knowledge and a kind of practice that is not usually made explicit in professional accounts of practice but is nonetheless critical for learning.

REPRESENTING AND CHARACTERIZING KNOWLEDGE STATES

The instructional design specifies the nature of the activities and forms of engagement that are intended to support learning. Each phase of instruc-

TABLE 2.1. Abbreviated Construct Maps

Construct	Initial Performance	Intermediate Performance	Highest Performance
Data Display (DaD)	DaD1(a): Create or interpret displays without relating to the goals of the inquiry.	DaD4(a): Display data in ways that use its continuous scale (when appropriate) to see holes and clumps in data.	DaD6(a): Discuss how general patterns or trends are either exemplified or missing from a subset of cases.
Meta-Representational Competence (MRC)	MRC1(a): Recognize that displays represent data, but misinterpret one or more elements of the display.	MRC3(a): Compare displays by indicating what each shows about the structure of the data.	MRC5(b): Coordinate different display types and formats to strengthen an argument.
Conceptions of Statistics (CoS)	CoS1(a): Use visual qualities of the data to summarize the distribution.	CoS3(d): Predict how a statistic is affected by changes in its components or otherwise demonstrate knowledge of relations among components.	CoS4(d): Predict and justify changes in a sampling distribution based on changes in properties of a sample.
Chance (Cha)	Cha1(a): View chance as being under the control of an agent.	Cha4(c): Recognize that an "unlikely" string of outcomes is possible and even expected over many repetitions of the event (e.g., 3H run in 10 coin tosses).	Cha6(b): Use sample space to estimate probability and/or relative frequency of a compound (aggregate) event.
Modeling Variability (MoV)	MoV1(a): Attribute variability to specific sources or causes.	MoV3(a): Use chance devise to represent variability.	MoV5(a): Judge model fit in light of variability across repeated simulation with the same model.
Informal Inference (InI)	InI1(a): Make a judgment or prediction according to personal experience or beliefs.	InI4(a): Make predictions based on regions of values such as clumps or holes, sometimes identified by particular values.	InI 7(b): Compare two objects (conditions, etc.) by considering the variations in the sampling distributions of their statistics.

tion is intended to support the development of knowledge about one or more components of statistical reasoning and to promote their coordination. The results of the design studies, in which these forms of knowledge were systematically supported, suggested a progression of cognitive landmarks of learning. For purposes of assessment design, these landmarks of learning were organized as *constructs*, each representing a dimension of knowledge development (Wilson, 2005). The constructs were informed by the results of the design studies but were also contested and refined over multiple iterations of assessment design and professional development. Currently, six constructs span the learning progression and represent the forms and states of knowledge that the progression attempts to support:

1. Data display (representational competence)
2. Meta-representational competence (comparing and considering tradeoffs among displays)
3. Conceptions of statistics (characterizing distributions by measuring their characteristics)
4. Chance (theoretical, empirical approaches to estimating probability)
5. Modeling variability (constructing models of chance variation)
6. Informal inference (model-based views of inference)

Cognitive landmarks for each construct are represented by *learning performances* that describe the form of cognitive activity that signals attainment of a particular milestone. Each learning performance is further illustrated by examples drawn from student work and performance in classrooms. Table 2.1 outlines the lower, mid-range, and highest cognitive milestones for each construct. These landmarks of learning were constructed by considering the results of multiple iterations of design studies.

DESIGNING AN ASSESSMENT SYSTEM

Based on this initial step of describing states of knowledge and intended developmental pathway for each construct, we designed an assessment system to provide formative feedback to teachers about ongoing student progress and to provide summative assessment for other purposes, largely related to anticipated demands of accountability and program evaluation. We followed Wilson's (2005) construct-centered design process, in which items were designed to measure specific learning performances for each construct, and across respondents, to elicit multiple levels of a construct. Item types included both multiple choice and constructed response. Student responses to constructed response items were first mined to develop scoring guidelines consisting of descriptions of student reasoning and examples of student work, both of which were aligned to specific levels of performance for each construct. The process involved multiple cycles of item design, scoring, fitting of items to psychometric models, and item revision/generation. For example, in light of student responses, we often refined scoring exemplars, including re-descriptions of student reasoning and inclusion of more or better examples of student work. We generated new items in response to gaps in coverage of the landmarks of student learning described by the constructs, discarded items that could not be repaired, and redesigned items to obtain better evidence of student reasoning in light of student responses. Occasionally, student responses to items could not be located on constructs but nonetheless appeared to indicate a reliably distinctive mode of reasoning. This led to revision of constructs and/or levels. For example, the construct of informal inference was amended to include an initial performance corresponding to student beliefs about deterministic sampling—that is, a belief that a sample can guide inference because it copies the values of subsequent samples.

```
                      Item 1   Item 2
               X |
              XX |                2.3
               X |
              XX |
2            XXX |
             XXX |
             XXX |
              XX |
            XXXX |    1.3
          XXXXXX |
1          XXXXX |
          XXXXXX |
        XXXXXXXX |
        XXXXXXXX |
       XXXXXXXXX |
     XXXXXXXXXXX |                2.2
0     XXXXXXXXXX |
         XXXXXXX |
       XXXXXXXXX |
        XXXXXXXX |
         XXXXXXX |
          XXXXXX |    1.2
-1         XXXXX |
            XXXX |
            XXXX |
              XX |                2.1
             XXX |
              XX |
-2            XX |
               X |    1.1
               X |
               X |
               X |
```

FIGURE 2.6. Wright map for two items of data display construct.

We illustrate our approach to analyzing empirical evidence about constructs and items with a sample of 847 middle school students in two states. Preceding the analysis reported, separate analyses suggested no significant scorer effects, so these are not included in the analysis. A partial-credit, one-dimensional item response (IRT) model was fit separately to each construct. The partial-credit model distinguishes among levels of each construct (Masters, 1982). For each item, we use threshold values (also called Thurstonian thresholds) to describe the empirical characteristics of the item. If an item has *k* scoring levels, then there are *k*-1 thresholds, one for each transition between scores. Each item threshold gives the ability level (in logits) that a student must obtain to have a 50% chance of scoring at the associated scoring category or above (as opposed to below). For example, suppose an item has three possible score levels (0, 1, and 2). In this case there will be two thresholds. Suppose that the first threshold has a value of -0.25 logits: This means that a student with an ability of -0.25 has an equal chance of scoring in category 0 compared to categories 1 and 2. If their ability is lower than the threshold value, then they have a higher probability of scoring in category 0; if their ability is higher than -0.25, then they have a higher probability of scoring in either category 1 or 2 (than 0). These thresholds are, by definition, ordered: In the given example, the second threshold value must be greater than -0.25.

Item thresholds can be graphically summarized in a Wright map, which simultaneously shows estimates for both the respondents and items on the same (logit) scale. Figure 2.6 is an example. On the left side, the distribution of person abilities is shown, where ability entails knowledge of the skills and practices described by the constructs. The person abilities have a roughly symmetric distribution. On the right side, two items and their threshold values are seen. Items 1 and 2 both have four possible score levels, and therefore three thresholds; 1.1 thus represents the threshold value between score categories 0 and 1 on Item 1, 2.2 represents the threshold value between score categories 1 and 2 on Item 2, and similarly for the remaining values. We can note that Item 2 is more difficult, since the threshold value for each set of corresponding score categories is higher.

Figure 2.7 depicts the Wright map obtained with the partial-credit model of student responses to items for one construct, Data Display (DaD). The construct includes five levels describing progress from initial focus on case values to the ability to interpret (and construct) displays that demonstrate characteristics of aggregates. The left panel shows the estimated person-ability distribution. The values are centered on 0 with the majority of ability estimates between -2 and 2 logits. The right panel displays threshold values for 22 items, with each item in its own column. The NL values along the bottom of the graph represent the first threshold value, that is, between scoring in category 0 (missing or completely off construct) and NL (reasoning

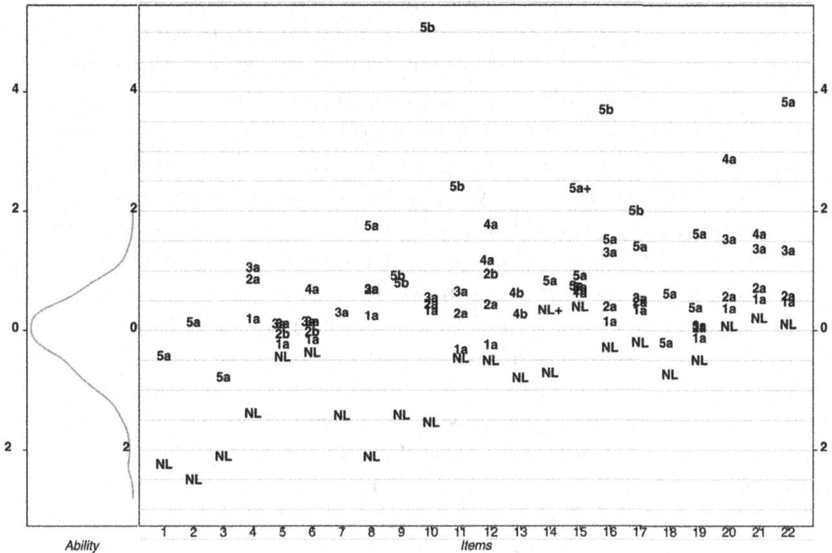

FIGURE 2.7. Wright map for data display construct, partial-credit model.

related to the item, but not sufficiently similar to any level defined by the construct). Moving up the graph, the points indicate the second and third thresholds, and so on.

The Wright map allows one to readily see whether there is empirical evidence for one form of construct validity: Items indicating the same level of performance on a construct ideally should be equally difficult. Thresholds indicating the same level of the construct should occupy similar regions on the map, which visually translate into horizontal bands. Inspection of the Wright map in Figure 2.7 suggests that for most items, these conditions are met. However, there are exceptions, and these exceptions suggest either that the theory of ordering should be modified or that the threshold observed includes more than just the construct level. For example, correct responses to the first three items indicate a high level (5) on the construct map, yet their item locations are below those of other items, corresponding to lower levels of the construct. Since these items were multiple choice rather than constructed response, the anomaly in ordering may relate primarily to respondents finding multiple choice items easier than constructed response items.

When examining items to determine whether they are performing in a satisfactory way, we use so-called *fit* statistics that indicate how much the performance of the item differs from what we expect from the estimated item parameters. In particular, we use the weighted fit statistic (sometimes known as *infit*), and the important issue is how much the statistic varies from 1.0. A value of 1.0 (and the region around that value) represents a good fit; values considerably above 1.0 represent situations in which the item is behaving in a way that is less consistent with the rest of the items than was expected. Values considerably below 1.0 represent situations in which an item is behaving in a way that is more consistent with the rest of the items than was expected. As with all statistics, we pay attention to both the effect size and the statistical significance of these fit statistics. In this case the effect size is decided using the value of the weighted fit statistic—historical experience has indicated that a range of .75 to 1.33 constitutes useful limits. The statistical significance is decided by calculating a confidence interval around 1.0 (which is accomplished using the ConQuest software; see, for example, Wu, Adams, Wilson, & Haldane, 2008). We usually reserve our attention to items that exceed the standard criteria in terms of *both* effect size and statistical significance. Of the 22 items included in the Wright map displayed in Figure 2.7, the fit statistic of only one item (15) was problematic. The infit value was 1.37 (exceeding 1.33), and it is outside of the confidence interval (0.73, 1.27). Upon further review, we decided to retain the item but will revisit this decision in the future after additional data are collected.

Multidimensional Analyses

After fitting unidimensional models and testing item performance for every construct, we fit the multidimensional random coefficients multinomial logit model (MRCML) to the student responses (Adams, Wilson, & Wang, 1997). This model estimates a profile of student ability, one for each construct, aligned on the same scale. As expected, the constructs are positively correlated, yet each is distinct, consistent with the learning trajectory described previously. In addition, the correlation structure improves the reliability of the person ability estimates because the estimate is now based on *both* the direct information from the items related to each dimension and information from the other dimensions (through the structure of the correlations). Thus, employing multidimensional models has a practical aspect as well, allowing the construction of shorter tests without the need to sacrifice reliability. The assessment system provides a practical means for indicating levels of knowledge valued by the learning progression and for measuring individual change.

DESIGNING PROFESSIONAL DEVELOPMENT

Professional development is an important component of the learning progression, both to prepare teachers to be effective pedagogically and to grasp the wider sense of statistics and data implied by the data modeling framework. To design and develop professional development, we worked with teachers (Year 1, n = 34; Year 2, n = 29) who had not participated in the original cycles of design. Seventeen teachers participated in two years of professional development. Our aims were to engage teachers to (1) further develop their knowledge of statistics and chance; (2) understand the design of the lessons, including knowledge of expected patterns of student thinking; and (3) employ the assessment system to interpret student responses and to use this information to guide instruction in classrooms. We did not take these elements as fixed, but rather as negotiated. We did so primarily by taking the position that the trajectories of student thinking outlined by the constructs were conjectures resulting from initial iterations of the instructional design and that the items and lessons designed to support student learning also enjoyed a provisional status. As we worked with teachers, we found it necessary to modify and even substantially transform material and conceptual arrangements that had served to coordinate the activity of the assessment and learning researchers.

One such transformation involved redesigning the construct maps to illuminate their relation to student talk and activity in classrooms, not solely in interviews or other settings. Teachers also wondered how items could be productively employed to foster transitions among the levels described by the constructs. Accordingly, we worked with teachers to revise the construct maps to meet these needs. We developed multimedia versions of the maps that included classroom-based, video examples of each learning performance. The initial examples were drawn from the design studies classrooms, but over time, teachers contributed episodes from their own classrooms. We also included video of formative assessment conversations that exemplified how teachers might employ items to support conceptual change. Video examples were, again, initially drawn from researcher design studies, but subsequent examples drew from the classrooms of participating teachers. Video annotations clarified how student talk, and activity corresponded to particular levels of one or more constructs. These annotations helped teachers view learning performances more dynamically as the formative assessment conversation unfolded.

A second transformation involved expression of the instructional trajectory as curriculum units designed to help teachers understand the intention of the designers (Davis & Krajcik, 2005; Schneider & Krajcik, 2002). Each instructional unit included (1) clear descriptions of the educative intention of each task or challenge posed to students, (2) student work demonstrating prototypical ways that students tend to reason and respond to each task or challenge, and (3) thought-revealing questions or ways to monitor student progress that were embedded in the course of classroom activity.

Assessment was positioned as a means for diagnosing student knowledge and as a tool for supporting the forms of conceptual development described by the learning progression. This effort included teachers' use of items in their classrooms and their subsequent interpretation of samples of the resulting student responses in light of the conjectured states of knowledge for a particular construct. Teachers often made suggestions that led to item revisions, initially focusing primarily on surface features of design, but over time, these revisions to items and to scoring guides increasingly reflected teachers' increasing knowledge of productive ways to elicit student reasoning.

Relations among teaching, learning, and assessment were explored during 13 day-long workshops over the course of two years. Each workshop typically began with opportunities for teachers to engage in a data modeling activity that was either identical to or closely related to one or more of those in the curriculum units. Engaging in this activity often raised mathematical issues and challenges, some of which closely resembled those we found with students during the initial design studies. The forms of thinking generated and employed during these investigations, usually with examples of student responses to the same or similar activities, were then compared to the descriptions of student thinking described by the construct maps. The workshops were also forums for considering how the instructional units could be employed as tools for learning: Teachers with experience teaching with the units often offered pedagogical advice to those who joined the project during its second year.

The assessment system was employed and modified as teachers engaged in two forms of practice. Teachers gave quizzes in their classrooms and brought samples of student work to score with the scoring guides developed by the assessment-learning teams. Scoring occasionally resulted in further modification to the scoring guides. Teachers also used items as instructional tools. To support use of items as instructional tools, we suggested construct-centered guidelines for classroom conversations, including (1) choice of student responses for sharing by considering their representation of different levels of the construct map, (2) staging fruitful comparisons among student responses by juxtaposing different levels of reasoning, and (3) engaging students whose responses were not chosen, perhaps by asking them to indicate which solution was closest to their own or to indicate how listening to others' solutions helped them reconsider their own position.

CLASSROOM PRACTICES AND THE LEARNING PROGRESSION

We observed and interviewed a group of teacher-workshop participants to characterize how teachers made use of the assessment system to support their students' learning (Kim, 2010a, 2010b). We observed ten grade 4–7 teachers (Year 1, eight teachers; Year 2, five teachers; three of the teachers were observed for two years). We conducted modified *teaching sets* (Simon & Tzur, 1999), consisting of a classroom observation and a follow-up interview with the teacher about his/her intentions regarding specific instructional moves and about his/her rationale for the organization of the formative assessment conversations (Black & Wiliam, 1998).

From ongoing analysis of these observations, we have tentatively identified four distinct forms of formative assessment practice. The first is one in which teachers appear to employ assessment items primarily as opportunities for gauging correctness of student performance. These classroom conversations featured an IRE (Initiate-Respond-Evaluate) pattern of classroom discourse (Mehan, 1979). The second form of formative assessment practice reflected an orientation toward using items to generate opportunities for participation. Here, classroom talk had a turn-taking, show-and-tell format, but we were unable to identify particular rationales for the selection of student presentations or their order. Follow-up conversations with teachers generally confirmed these impressions. The third form of formative assessment practice focused on intentionally eliciting student responses that illuminated the diversity of forms of reasoning identified by the construct maps and scoring guides. This form of classroom conversation was typically formatted as a succession of demonstrations, with time allocated for classroom questions and follow-up. The fourth form of practice included both teachers' intentionally eliciting student responses, guided by the construct maps, and teachers' efforts to deliberately compare and contrast different levels of reasoning. This characterization of teacher practice afforded analysis of prospective trajectories of change in teacher practices. We report a comparative analysis of two teachers to illustrate some of these transitions.

The first teacher, Rana (teachers' names are pseudonyms), had a bachelor's degree in mathematics and a master's degree in education. Just beginning her teaching career, she taught seventh-grade mathematics and participated in the data modeling workshop for two years. During the first year of observation, Rana often conducted formative assessments by reading an item to the entire class. Students worked in small groups and shouted out their answers. Rana responded to affirm correct responses. The conversation format was that of IRE.

By the end of the second year of the project, however, Rana's formative assessment practices had become substantially more nuanced. She monitored students' thinking as they worked on assessment items individually and chose students to share solutions based on how well their reasoning exemplified the different levels of learning performances described by the construct guiding the assessment. She often revoiced (O'Connor & Michaels, 1993) and summarized what students said (e.g., "Is that what you are saying?") to ensure that student contributions were heard by the entire class. Many of the student responses Rana selected in the second year would not have received classroom time or attention during the first year. Rana apparently used the construct map in mediating this change in her practice, as she recalled during an interview:

> Earlier I didn't think that you really needed to talk about wrong answers because we don't want wrong answers, so we don't wanna talk about them. So it showed me that wrong answers definitely have a place—or misguided answers—because they're really just basic understanding. It's the beginning blocks of them being able to understand all these different mathematical concepts.

The second teacher, Maggie, had a bachelor's degree in science and a middle school endorsement in math and history. More experienced than Rana, she had been teaching mathematics for eight years (fourth grade for two years, fifth and sixth grades for six years). Although she joined the project in its second year, her initial approach to the formative assessments was oriented toward the construct maps and video exemplars, perhaps because by this point in the project, these objects were now in common use. Maggie first administered items in a class quiz. She used the scoring guides to identify classes of student responses, with an eye toward conducting a formative conversation about typical patterns of response, "Because several of my kids did it like this. So I wanted to get those kids up there to talk about their way." She also identified unique forms of reasoning: "[I] tried to get the ones who did it different. Like he had done it totally different from everybody." As students presented their solutions, Maggie drew attention to important mathematical features of the response. But unlike Rana, Maggie also invited students to compare and contrast different forms of student reasoning. The video-exemplar construct map mediated this form of practice, especially its illuminations of formative assessment practices. For example, Maggie noted that when she watched video examples of assessment, she paid particular attention to how the teacher orchestrated the classroom conversation:

> It was what he [he teacher] was saying in response to the kids, how he was pushing them farther, how he gets them to talk to each other and bounce off of each other and when they disagree. I more like to watch those things to get what the teacher says and how they respond to what the kids have said about it, just to get it to go farther 'cause my biggest fear is they answer the question and I go, "Okay, next." You know, 'cause I have nowhere to go with it and I'm like, "Okay, yeah," you know.

CHARACTERIZING LEARNING TRAJECTORIES

Having engaged in this extensive spadework, the project's learning sciences and psychometric teams are currently exploring means of representing and

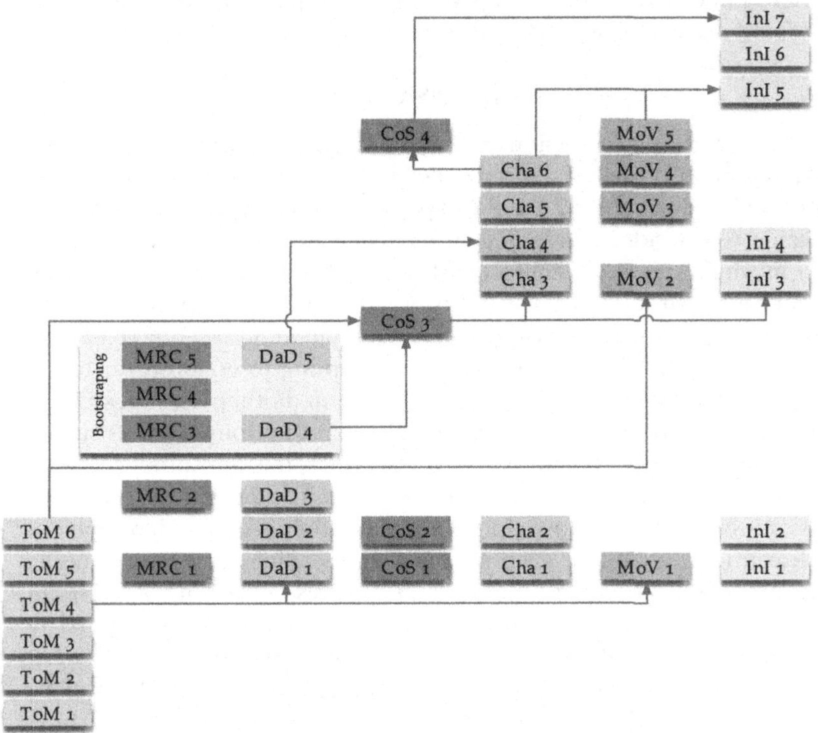

FIGURE 2.8. Prospective conceptual development pathways. Note: Prospective pathways of conceptual development in the outcome space defined by constructs and learning performances.

testing pathways in the space defined by the outcome measures. These prospective pathways relate levels of the constructs according to the trajectory envisioned and outlined earlier. Figure 2.8 suggests one such model, which assumes that students are taught by teachers whose practices offer sustained support for the forms of learning envisioned in the trajectory. The arrows in Figure 2.8 depict causal relations among states of the constructs, thus allowing for more refined articulations of the conceptual trajectories envisioned by learning progression. The ToM (Theory of Measurement) construct in the figure articulates learning performances in measurement that are taken as resources by the learning progression in data and statistical reasoning. Bootstrapping refers to co-origination of representational and meta-representational competencies. The work of articulating and contesting models like these is ongoing and is made feasible by prior articulation of states of knowledge, means of pedagogical support, and reliable indicators of student progress.

DISCUSSION

Although learning progressions include prospective pathways of conceptual development, this view of progressions tends to privilege one of the strands outlined in Figure 2.1 at the expense of others. We believe it is more profitable to consider a learning progression as a *trading zone* (Galison, 1997) in which different realms of educational practice intertwine, much as a cable is constructed. We have considered strands of learning theory, assessment, instruction, and professional development as essential elements of the learning progression that we are attempting to establish for teaching and learning statistics. We have not highlighted other relevant strands, such as software design and development, or institutions, such as school districts and national standards boards, in which this and other learning progressions are embedded. Each of these communities has its own set of issues, questions, and history, and at different points of the development of a learning progression, each may take the leading role. For example, we described a process that began with an analysis of learning and a series of classroom design studies that resulted in alterations of the design of instruction, especially the forms of pedagogical support that appeared necessary for student progress. As instruction stabilized over repeated iterations of design, we turned more attention to assessment, recasting the learning trajectory into construct maps and designing items by repurposing questions that had been posed to students in classrooms or in clinical interviews. The system of assessment was informed by broader institutional concerns about accountability, as well as by the need for teaching to be informed by information about students' current states of knowledge. The participants in each team continued to work within the bounds of their own profession, but the learning progression was a site for cross-boundary work that was never completely interior to any of these disparate worlds of educational practice.

Boundary crossings among learning, assessment, and teaching were facilitated by the development of boundary objects and by brokers who participated in two or more of these communities (Lehrer, 2007). Boundary objects are material and conceptual systems that distinct communities can use to coordinate their work because the boundary objects can be shared yet allow for autonomous work within each community (Bowker & Star, 1999; Star & Griesemer, 1989). In this work, the most prominent boundary objects were the construct maps, although the assessment items, scoring guides, and curriculum units also played significant roles.

For the assessment team, the construct map functioned as an explicit declaration of what was worth learning, and as a set of knowledge outcomes to be sought. This declaration then served as the framework for the development of the assessment system, and its use helped the assessment team hold item design and item calibration accountable to the constructs.

For teachers, the constructs functioned as heuristics that suggested pathways of student development, but to understand how to support student learning, the teachers required examples of what students might say and do at each state of knowledge. Teachers also requested examples of teaching practice that would support conceptual change. To amend the construct map to include video exemplars of student reasoning and teaching practices, the learning sciences team worked with teachers to identify video clips appropriate for each purpose. Initially, the relationship between video clips and constructs was ill structured, but as teachers used the construct maps as guides to their practice, the nature of the necessary revisions became clearer and hence the video clips became more well structured in relation to the construct map (see Star, 2010, re: *tacking* and boundary objects).

For the learning science researchers, the constructs presented significant challenges for encapsulating learning as a series of performances and for choosing cases that could be considered examples of these performances. The learning researchers viewed the construct maps as a locus for further research about transformations of teaching practice, but the linear and ordered format of the construct map did not conform to any theory of learning validated by the learning community. That is, for them, the construct maps had the status of the Bohr model in contemporary chemistry—good enough to get some work done by coordinating the activity of learning researchers, teachers, and assessment specialists, but clearly an inadequate description of a learning process. For example, Lehrer, Kim, and Jones (2011) describe how a classroom activity structure, a measure review, supported student development of concepts of statistics, thus revealing the work of learning not visible in the learning performances of the corresponding construct.

What was traded in the zone established by the learning progression were approximations of practice in each community that were sufficient for generating and sustaining the learning progression in middle school classrooms. The boundary objects did not serve merely to coordinate the organization of activity in these realms of professional practice. They also destabilized the organization of activity, and this role, too, supported the progress of the project. For example, levels of performance with respect to constructs were revised in light of student responses to items, and new design studies were initiated to fill in gaps in our understanding of student thinking that were suggested by these student responses. As the work in the assessment system developed, the learning researchers began to recast conceptual change in terms of the space of outcomes defined by the assessments, as signaled by Figure 2.8. This created a press for innovation in forms of modeling by the project's psychometric team (e.g., Diakow, Irribarra, & Wilson, 2011; Schwartz, Ayers, & Wilson, 2010). These innovations also furthered the capacities of the psychometric community, for which this project was but an exemplar.

For some of the participating teachers, images of their own professional practice began to include authoring representations of their classroom practices and objectifying these practices in the video exemplar construct map. This, too, was transformative of teaching practices, spurring some teachers to seek new branches in their own professional trajectories, primarily assuming new roles as math coaches or as math specialists in their districts. Many teachers' views of the goals and practice of assessment changed in ways illustrated by the cases presented. Hence, as the work progressed, members of each community were challenged to transform themselves in ways that would serve the learning progression. And, at the boundaries, contests about the intelligibility, transparency, and ultimate worth of the constructs were commonplace, though not so vigorous as to disrupt the entire enterprise.

The work of validating our proposed learning progression appears to more closely resemble the activity and standards around which theories and programs of research (Lakatos, 1976) are judged than it does a traditional psychometric or falsification model of validation. That is, our learning progression centers around a program of professional work that must be intelligible to multiple audiences, should generate successive rounds of design, and must ultimately be of consequence to students' learning to represent and model data and uncertainty. The last aim clearly has implications that extend well beyond the borders of schooling. Perhaps this quality of boundary spanning should be included as educators consider topics worthy of extended conceptual development and, hence, of the design of a learning progression.

AUTHOR NOTE

This research was supported by the National Science Foundation, REC 0337675, and by the Institute of Education Sciences, U.S. Department of Education, through Grant R305K060091 to Vanderbilt University. The opinions expressed are those of the authors and do not represent views of either the U.S. Department of Education or the National Science Foundation. We thank Cesar Delgado, Leona Schauble, and the editors for their constructive critiques of earlier versions of this work.

REFERENCES

Adams, R. J., Wilson, M., & Wang, W. (1997). The multidimensional random coefficients multinominal logit model. *Applied Psychological Measurement, 21*(1), 1–23.

Boaler, J. (2002). *Experiencing school mathematics.* Mahwah, NJ: Lawrence Erlbaum Associates.

Berlinski, D. (2000). *The advent of the algorithm.* New York, NY: Harcourt.

Black, P., & Wiliam, D. (1998). Assessment and classroom learning. *Assessment in Education, 5*(1), 7–74.

Bowker, G., & Star, S. L. (1999). *Sorting things out: Classification and its consequences.* Cambridge, MA: MIT Press.

Cobb, G. W., & Moore, D. S. (1997). Mathematics, statistics, and teaching. *The American Mathematical Monthly, 104*(9), 801–823.

Cobb, P. (1999). Individual and collective mathematical learning: The case of statistical data analysis. *Mathematical Thinking and Learning, 1*(1), 5–44.

Cobb, P., Confrey, J., diSessa, A., Lehrer, R., & Schauble, L. (2003). Design experiments in education research. *Educational Researcher, 32*(1), 9–13.

Davis, E. A., & Krajcik, J. S. (2005). Designing educative curriculum materials to promote teacher learning. *Educational Researcher, 34*(3), 3–14.

Diakow, R., Irribarra, D. T., & Wilson, M. (2011, April). *Analyzing the complex structure of a learning progression: Structured construct models.* Paper presented at the annual meeting of the American Educational Research Association, New Orleans, LA.

diSessa, A. (1995). Epistemology and systems design. In A. diSessa, C. Hoyles, R. Noss, & L. D. Edwards (Eds.), *Computers and exploratory learning* (pp. 15–29). Berlin, Germany: Springer Verlag.

diSessa, A. (2004). Metarepresentation: Native competencies and targets for instruction. *Cognition and Instruction, 22*(3), 293–331.

Galison, P. (1997). *Image and logic. A material culture of microphysics.* Chicago, IL: University of Chicago Press.

Garfield, J., DelMas, R. C., & Chance, B. (2007). Using students' informal notions of variability to develop an understanding of formal measures of variability. In M. C. Lovett & P. Shah (Eds.), *Thinking with data* (pp. 117–147). Mahwah, NJ: Lawrence Erlbaum Associates.

Greeno, J. G., & Hall, R. (1997). Practicing representation: Learning with and about representational forms. *Phi Delta Kappan, 78*(1), 361–367.

Gresalfi, M. S. (2009). Taking up opportunities to learn: Constructing dispositions in mathematics classrooms. *Journal of the Learning Sciences, 18*(3), 327–369.

Kim, M.-J. (2010a, April). A case of collaboration between researchers and teachers mediated by boundary objects. In R. Lehrer & M. Wilson (Chairs), *Assessing a multidimensional learning progression: Psychometric modeling and brokering professional development.* Symposium conducted at the annual meeting of the American Educational Research Association, San Diego, CA.

Kim, M.-J. (2010b, April). Collaboration with teachers: Brokering the assessment system. In R. Lehrer & M. Wilson (Chairs), *Constructing a multidimensional learning progression for data modeling: Design studies, psychometric modeling and brokering professional development.* Symposium conducted at the meeting of the National Council of Teachers of Mathematics Research Presession, San Diego, CA.

Knorr-Cetina, K. (1999). *Epistemic cultures: How the sciences make knowledge.* Cambridge, MA: Harvard University Press.

Konold, C. (2007). Designing a data analysis tool for learners. In M. C. Lovett & P. Shah (Eds.), *Thinking with data* (pp. 267–291). Mahwah, NJ: Lawrence Erlbaum Associates.

Konold, C., & Lehrer, R. (2008). Technology and mathematics education: An essay in honor of Jim Kaput. In L. D. English (Ed.), *Handbook of International Research in Mathematics Education Second Edition* (pp. 49–72). New York, NY: Routledge.

Konold, C., & Miller, C. D. (2005). TinkerPlots: Dynamic data exploration. [Computer software]. Emeryville, CA: Key Curriculum Press.

Konold, C., & Pollatsek, A. (2002). Data analysis as the search for signals in noisy processes. *Journal for Research in Mathematics Education, 33*(4), 259–289.

Lakatos, I. (1976). *Proofs and refutations: The logic of mathematical discovery.* Cambridge, MA: Cambridge University Press.

Lehrer, R. (2007, April). *Collaboration at the boundaries: Brokering learning and assessment improves student achievement.* Symposium presented at the annual meeting of the American Educational Research Association, Chicago, IL.

Lehrer, R. (2009). Designing to develop disciplinary knowledge: Modeling natural systems. *American Psychologist, 64*(8), 759–771.

Lehrer, R., & Kim, M.-J. (2009). Structuring variability by negotiating its measure. *Mathematics Education Research Journal, 21*(2), 116-133.

Lehrer, R., Kim, M.-J., & Jones, S. (2011). Developing conceptions of statistics by designing measures of distribution. *International Journal on Mathematics Education (ZDM), 43*(5), 723–736.

Lehrer, R., Kim, M.-J., & Schauble, L. (2007). Supporting the development of conceptions of statistics by engaging students in modeling and measuring variability. *International Journal of Computers for Mathematics Learning, 12*(3), 195–216.

Lehrer, R., & Romberg, T. (1996). Exploring children's data modeling. *Cognition and Instruction, 14*(1), 69–108.

Masters, G. N. (1982). A Rasch model for partial credit scoring. *Psychometrika, 47*(2), 149–174.

Mehan, H. (1979). *Learning lessons: Social organization in the classroom.* Cambridge, MA: Harvard University Press.

Moore, D. S. (1990). Uncertainty. In L. Steen (Ed.), *On the shoulders of giants: New approaches to numeracy* (pp. 95–137). Washington, DC: National Academy Press.

National Research Council. (2006). *Systems for state science assessments.* (Committee on Test Design for K–12 Science Achievement, M. R. Wilson & M. W. Bertenthal, Eds.). Board on Testing and Assessment, Center for Education. Division of Behavioral and Social Sciences and Education. Washington, DC: The National Academies Press.

Nersessian, N. J. (2008). *Creating scientific concepts.* Cambridge, MA: MIT Press.

O'Connor, M. C., & Michaels, S. (1993). Aligning academic task and participation status through revoicing: Analysis of a classroom discourse strategy. *Anthropology & Education Quarterly, 24*(4), 318–335.

Petrosino, A. J., Lehrer, R., & Schauble, L. (2003). Structuring error and experimental variation as distribution in the fourth grade. *Mathematical Thinking and Learning, 5*(2&3), 131–156.

Piaget, J. (1970). *Genetic epistemology.* New York, NY: Norton.

Schneider, R. M., & Krajcik, J. (2002). Supporting science teacher learning: The role of educative curriculum materials. *Journal of Science Teacher Education, 13*(3), 221–245.

Schwartz, R., Ayers, E., & Wilson, M. (2010, April). *Modeling a multi-dimensional learning progression.* Paper presented at the annual meeting of the American Educational Research Association. Denver, CO.

Simon, M. A. (1995). Reconstructing mathematics pedagogy from a constructivist perspective. Journal for Research in Mathematics Education, 26(2), 114–145.

Simon, M. A., & Tzur, R. (1999). Explicating the teacher's perspective from the researchers' perspectives: Generating accounts of mathematics teachers' practice. *Journal for Research in Mathematics Education, 30*(3), 252–264.

Star, S. L. (2010). This is not a boundary object: Reflections on the origin of a concept. *Science, Technology, & Human Values, 35*(5), 601–617.

Star, S. L., & Griesemer, J. (1989). Institutional ecology, 'translations,' and boundary objects: Amateurs and professionals on Berkeley's museum of vertebrate zoology, 1907–39. *Social Studies of Science, 19*(3), 387–420.

Thompson, P. W., Liu, Y., & Saldanha, L. A. (2007). Intricacies of statistical inference and teachers' understanding of them. In M. C. Lovett & P. Shah (Eds.), *Thinking with data* (pp. 207–231). Mahwah, NJ: Lawrence Erlbaum Associates.

Torok, R., & Watson, J. (2000). Development of the concept of statistical variation: An exploratory study. *Mathematics Education Research Journal, 12*(2), 147–169.

Watson, J. M. (2006). *Statistical literacy at school: Growth and goals.* Mahwah, NJ: Lawrence Erlbaum Associates.

Wild, C. J., & Pfannkuch, M. (1999). Statistical thinking in empirical enquiry. *International Statistical Review, 67*(3), 223–248.

Wilson, M. (2005). *Constructing measures.* Mahwah, NJ: Lawrence Erlbaum Associates.

Wu, M., Adams, R.J., Wilson, M., & Haldane, S. (2008). ACER*ConQuest* 2.0 [Computer Software]. Hawthorn, Australia: ACER.

Zaojewski, J. S., & Shaughnessy, J. M. (2000). Mean and median: Are they really so easy? *Mathematics Teaching in the Middle School, 5*(7), 436–440.

CHAPTER 3

EQUIPARTITIONING, A FOUNDATION FOR RATIONAL NUMBER REASONING

Elucidation of a Learning Trajectory

Jere Confrey, Alan P. Maloney,
Kenny H. Nguyen, and André A. Rupp

ABSTRACT

This chapter discusses research on learning trajectories for rational number reasoning, illustrated by the development of a learning trajectory for equipartitioning, a construct grounded in student generation of fair shares. This work is set in a larger context of development and validation of learning trajectories as the underlying framework for diagnostic assessment approaches to formatively monitoring and instructionally supporting the development of student reasoning about major mathematical concepts.

Learning Over Time: Learning Trajectories in Mathematics Education, pages 61–96.

LEARNING TRAJECTORIES

An orientation toward understanding the development of student knowledge over time spurred the idea of a *learning trajectory*. Simon's (1995) introduction of a *hypothetical learning trajectory*, defined as "the learning goals, the learning activities, and the thinking and learning in which students might engage" (p. 133), described how teachers enter the teaching process with hypotheses of how they expect their students to reason on mathematical tasks, based on practice and research. For Simon, hypothetical learning trajectories are a framework to help anticipate the evolution of student thinking, and not a direct recipe for instruction.

Clements and Sarama (2004) modified the definition, stating, "A complete hypothetical learning trajectory includes all three aspects: the learning goal, developmental progression of thinking and learning, and sequence of instructional tasks" (p. 84). They further explained:

> Integration [of the learning goal, developmental progression, and sequence of tasks] can produce novel results, even within the local theoretical fields of psychology and pedagogy. The enactment of an effective, complete learning trajectory can actually alter developmental progressions or expectations previously established by psychological studies because it opens up new paths for learning and development. (p. 84)

Clements, Wilson, and Sarama (2004) remind us that any constructed trajectory is hypothesized to be a productive route but is not necessarily the most productive route for all students.

Other researchers use a variety of terms and metaphors to convey similar meanings. For example, Confrey (2006) defined a *conceptual trajectory* as a possible path students can navigate during any particular set of instructional episodes, and a conceptual corridor as all likely conceptual trajectories. Figure 3.1 illustrates these concepts. This figure indicates that student thinking begins with their prior knowledge and how it connects to a problematic as an invitation into the new ideas. The task sequence sets the borders of the corridor. Students travel the corridor in different ways but within the corridor are landmarks—important ideas they must encounter—and obstacles or predictable patterns of errors or misconceptions they typically navigate to make progress through the space.

Battista (2009) refers to a cognitive terrain consisting of several plateaus of knowledge from informal conceptualization and reasoning to formal mathematical reasoning with several possible learning trajectories woven through the plateaus. He calls this a *levels model* for a topic.

In our own research group, called DELTA (Diagnostic E-Learning Trajectories Approach), we have further refined the definition of a learning trajectory as follows:

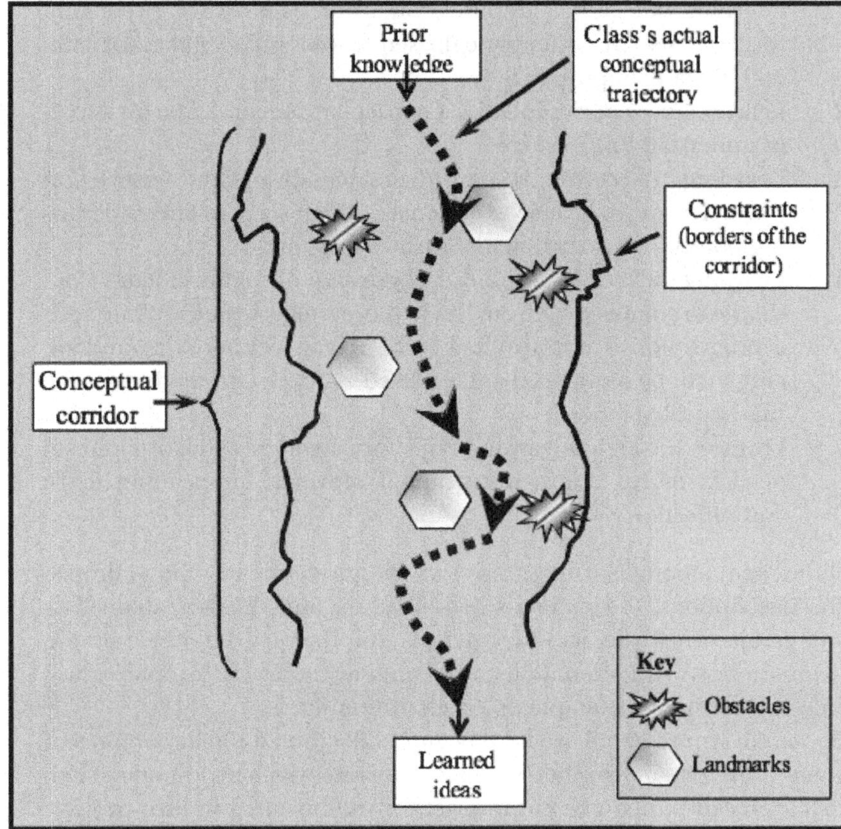

FIGURE 3.1. Conceptual corridor, showing a learning trajectory with conceptual landmarks and obstacles (Confrey, 2006)

> A researcher-conjectured, empirically supported description of the ordered network of constructs a student encounters through instruction (i.e., activities, tasks, tools, forms of interaction and methods of evaluation), in order to move from informal ideas, through successive refinements of representation, articulation, and reflection, towards increasingly complex concepts over time. (Confrey, Maloney, Nguyen, Mojica, & Myers, 2009)

Learning trajectories have now themselves progressed to the status of a tool that can both scaffold instruction and be enacted through instruction in the classroom, as other chapters in this volume attest. Across the different meanings of learning trajectories, five common assumptions have emerged:

1. Learning trajectories are based on synthesis of existing research, further research to complete the sequences, and a validation method based on empirical study.
2. Learning trajectories identify a particular domain and a goal level of understanding or target.
3. Learning trajectories recognize that children enter instruction with relevant yet diverse experiences, which serve as effective starting points for instruction and further learning.
4. Learning trajectories assume a progression of cognitive states from simple to complex (or from less to more sophisticated). While such a progression is not assumed to be linear, neither is it random, and it can be sequenced and ordered as "expected tendencies" or "likely probabilities."
5. Progress through a learning trajectory assumes a well-ordered set of tasks (curriculum), instructional activities, interactions, tools, and reflection.

Differences among learning trajectory theorists reside in the grain size of the descriptions, the methods of validation, and theories about how learning trajectories interact with curricula. Another key difference among learning theorists is how students move among the levels, a topic we will address further in our example on equipartitioning.

In our current research, we have begun to document the importance of learning trajectory approaches as a means to assist teachers in understanding what precedes their current instruction and how that in turn projects into subsequent instruction (see also Chapters 1, 2, and 8, this volume). This encourages teachers to engage in both vertical and horizontal alignment—coordination both within and across grades. For example, Wilson (formerly a member of our research group) conducted a 12-week study with elementary teachers to investigate how teachers might come to use learning trajectories within their practice (Wilson, 2009; Chapter 8, this volume). He observed and analyzed classroom practices of 10 second-grade teachers (from a group of 33 K–2 teachers who learned about the equipartitioning learning trajectory through 20 hours of professional development). As a result of his design experiment, Wilson identified five areas in which learning trajectories can have an impact on teachers' practice:

1. Building models of students' thinking
2. Judging the relative difficulty of instructional tasks
3. Analyzing students' work
4. Interacting with students during classroom instruction
5. Relating learning trajectories to curriculum.

In another study of 46 college students over eight weeks, Mojica (2010) demonstrated that in working with a learning trajectory, preservice elementary teachers who initially reported only on children's appearance of engagement with tasks (enthusiasm, frustration) and hastily labeled students' behaviors summarily as right or wrong developed abilities to draw more precise distinctions, relating children's discourse to more advanced uses of mathematical language and locating their evidence in actions the children took with particular tasks. Overall, in both studies, in-service and preservice teachers alike shifted from making assumptions about what children could *not* learn at a grade level to an orientation toward anticipating, recognizing, and facilitating what students *could* learn, with specific knowledge of what to expect. Both cases documented improved teacher content knowledge of the topic in question.

We outline below how we went about constructing a learning trajectory, including proficiency levels, outcome spaces and task classes, and corresponding assessment items. We illustrate the development of the equipartitioning learning trajectory, one of seven learning trajectories under investigation, and consider the implications of statistical analysis of a field test of the assessment items for the validity of the learning trajectory.

OVERVIEW OF RESEARCH PROGRAM

The overall aim of the DELTA project is to develop learning trajectories for rational number reasoning for grades K–8, along with diagnostic assessments based on those learning trajectories, and deploy the learning trajectories and diagnostic assessments as tools for instructional guidance. Instructional guidance in this case takes the form, first, of learning trajectories as a theory of student mathematical learning—a research-based characterization of student cognitive development of conceptual understanding—and a scaffold for professional development and preservice education about students' progression through the "big ideas" in mathematics. Second, the instructional guidance, in the context of the DELTA project, will take the form of learning trajectory-based diagnostic assessments and subsequent rapid feedback generated from them, for formative use by teachers and students in classrooms.

Our focus on learning trajectories and new models of diagnostic assessment derive from an overarching goal of supporting improvement and coherence of the "instructional core" in mathematics (Elmore, 2002). The standards and accountability movement of the 1990s, and in particular the 2001 reauthorization of the Elementary and Secondary Education Act (No Child Left Behind), put a primary focus on the two "bookends" of school accountability, standards, and high-stakes assessment (Figure 3.2). In many ways, the No Child Left Behind Act, implemented with an overemphasis on particular types of summative assessment, deflected attention from—and

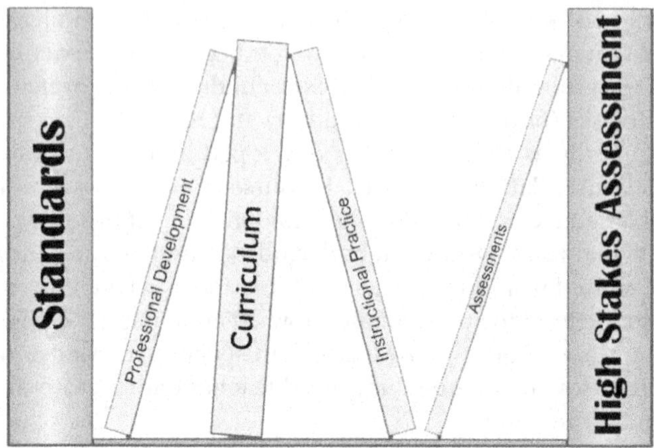

FIGURE 3.2. Standards and high stakes assessments as bookends of the accountability system, with components of classroom activity stacked between

in some ways actually undermined support for—a rich, coherent classroom learning environment.

Figure 3.3 provides an alternate orientation to the accountability bookends, placing the instructional core squarely at the center of classroom education. This figure reflects some of the lessons learned during the No Child Left Behind era, especially the value of disaggregating student achievement data as well as the disruptive effects of imbalanced focus on the accountability bookends. Figure 3.3 brings to the fore the interplay among curriculum (both as designed and as implemented), instructional practices, and professional development, all undergirded by a student learning-based framework of learning trajectories, and all interlinked by several kinds of classroom assessment practices, with iterative cycles of feedback among all these components.

Iterative design of the learning trajectory and assessment items is a central part of this model, in order that student responses to assessment items and tasks can be used to infer students' progress in content-area reasoning and provide the basis for instructional guidance to teachers. As a guide for linking learning trajectories to diagnostic assessment approaches, we have adopted the measurement tradition of *evidence-based assessment* (Mislevy, Steinberg, & Almond, 2003; Mislevy, Steinberg, Almond, & Lukas, 2006; Pellegrino, Chudowsky, & Glaser, 2001; Wilson, 2005). In this tradition, assessment and measurement specialists collaborate with content specialists to design assessments that are grounded in deep and careful study of student thinking over time.

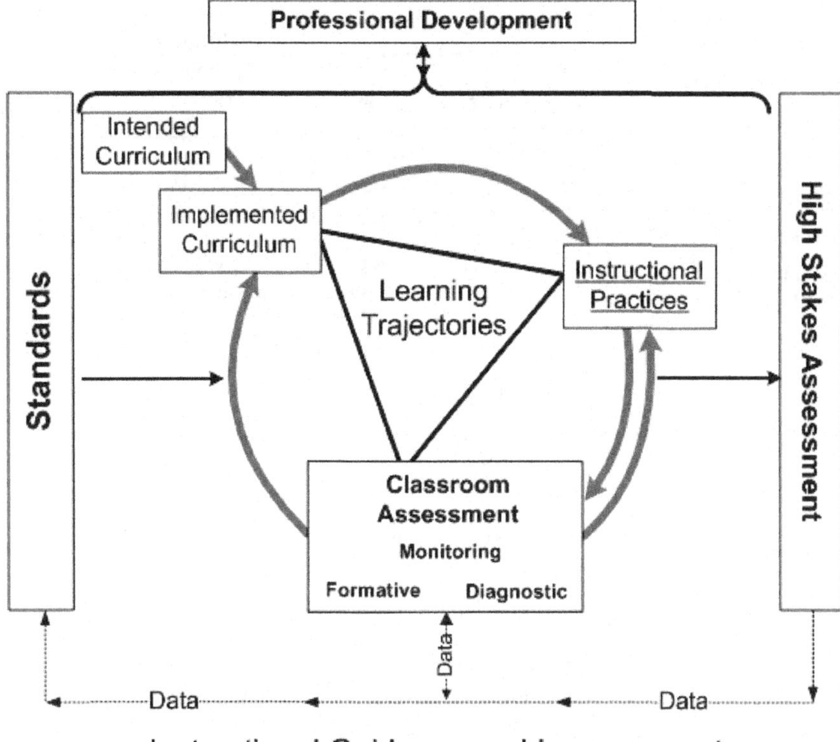

FIGURE 3.3. Model of the instructional core of classroom education, based on learning trajectories and rich coherent feedback-driven instruction, bounded by accountability bookends (Confrey & Maloney, 2012)

METHODOLOGY: FROM SYNTHESIS TO DIAGNOSTIC LEARNING PROFILES

The methodology depicted in Figure 3.4 represents the systematization of our experience in developing a learning trajectory for construct of rational number reasoning, *equipartitioning*, and of diagnostic assessment approaches for this construct. This methodology is in turn used to guide the development of learning trajectories and diagnostic assessments for other rational number reasoning constructs. It incorporates the methods of synthesis, clinical interviewing, teaching experiment, and methods of building and analyzing assessments (Confrey & Maloney, 2012; Maloney & Confrey, 2010). Portions of this methodology are shared by other learning trajectory researchers (see Chapters 1, 2, and 4, this volume).

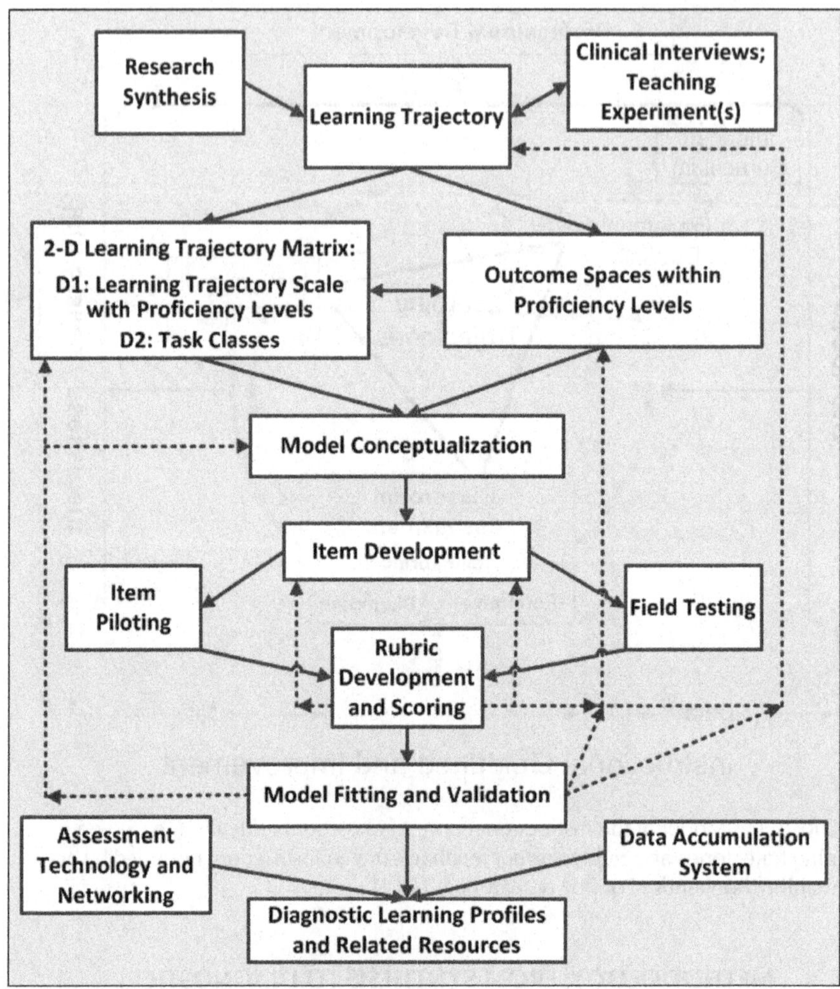

FIGURE 3.4. Methodology for developing learning trajectories and diagnostic assessments
Note: Flow chart of research method for developing learning trajectories and learning trajectory-based diagnostic assessment. Dashed lines indicate feedback and revision cycles (from Confrey & Maloney, 2012).

Conjectured Learning Trajectories for Rational Number Reasoning

To facilitate research syntheses in rational number, Confrey (2008) mapped seven major rational number concept areas as a set of conjectured learning trajectories (Confrey, Maloney, Nguyen, Wilson, & Mojica, 2008): equipartitioning, ratio and rate, fraction-as-number, similarity and scaling,

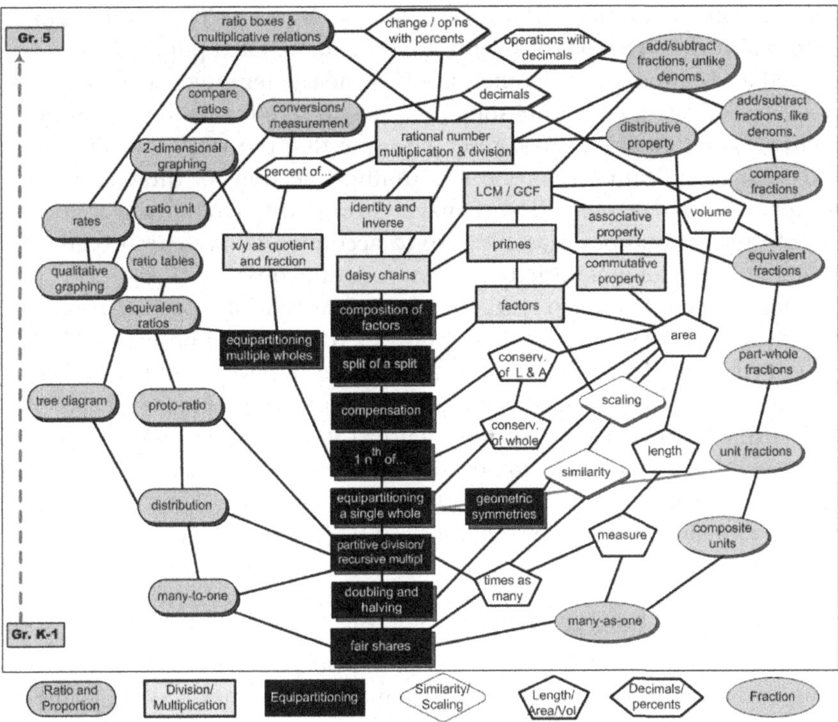

FIGURE 3.5. Map of rational number reasoning concepts
Note: Learning trajectories for major concept areas are indicated with different shapes and shading; key at bottom of FIGURE (after Confrey, 2008, and Confrey et al., 2009). See www.gismo.fi.ncsu/Chap3Fig3.5 for a magnifiable color version.

multiplication and division, length and area, and decimals and percents (Figure 3.5).

Initial (Iterative) Empirical Support: Articulation of Learning Trajectory Levels Through Research Synthesis and Clinical Interviews

In the case of rational number reasoning, there is an extensive body of research, including much that was part of the Rational Number Project, a network of research projects funded by the National Science Foundation beginning in 1979 (Behr, Cramer, Harel, Lesh, & Post, 2011). Our research team conducted extensive searches of the literature and directly solicited contributions from many researchers. We constructed a database of more than 625 journal articles, book chapters, and conference proceedings on rational number reasoning (Confrey, Maloney, & Nguyen, 2008), which can be found at http://gismo.fi.ncsu.edu/RNRdatabase. Each publication was described and summarized. Publicly available assessment items from those works were collected. Each published work was assigned to a single learning trajectory, based on the central research topic that the work investigated and to avoid duplicating references in multiple sections of the database. However, a particular research work may inform more than one major construct of rational number reasoning, so to accommodate research syntheses, the database can be queried for numerous keywords.

We begin by conducting a synthesis of the research for a particular construct, drawing heavily on the description of research synthesis provided by Cooper (1998). In particular, a *synthesis* is not intended to be merely a review of literature but instead to propose "overarching schemes that help make sense of many related but not identical studies" (p. 12). We regard research synthesis as a means to advance mathematics education research by bringing together cumulative results "more complex than any single study...to explain higher-order relations" (p. 13), with the explicit possibility that synthesis can itself lead to the identification of new ideas. This is particularly the case when a field (such as mathematics education in general, or education and assessment in rational number in particular) has accumulated a rich research base, and technological, policy, or societal trends impinge on the science to make new demands or open new ways to frame previously established norms.

Our research syntheses draw on disparate lines of research inquiry (mathematics education, cognitive psychology, and others) and different age ranges of students, focusing on grades Pre-K through 8. As an outcome of the research synthesis (based on our interpretation from the research literature), we conjecture a progression of student reasoning about related concepts, from less sophisticated to more sophisticated, framed as hierarchical cognitive levels, and foreshadowing subsequent hypotheses about relative difficulty of assessment items corresponding to the proficiency levels.

After we develop the research synthesis, we conduct structured clinical interviews (Ginsburg, Kossan, Schwartz, & Swanson, 1983; Opper, 1977; Piaget, 1970) to provide further insight into student reasoning in particular settings and topics. These interviews are aimed at enriching our understanding of student behaviors, representations, and utterances related to different specific contexts and cognitive challenges, and thereby to fill in gaps in the levels of the initial trajectory and establish the boundaries of a learning trajectory. The clinical interview is an essential tool for this kind of inquiry, making visible a broad range of student reasoning and mathematical cognitive behaviors.

Small-group teaching experiments are also employed in this work. These are important for exploring students' approaches to novel problem situations for which they require some instructional background or review, in order to appraise cognitive steps they may take at the edge of their current understanding. In our teaching experiments with multiple students, we also encourage student–student discourse, facilitating further observation and understanding of the range of behaviors and reasoning students may exhibit when exploring new mathematical reasoning territory about a particular topic.

Creation of Learning Trajectory Matrix

Construction of learning trajectories and their representations is a highly iterative process. We attempt throughout to identify missing elements and to fill in gaps in the initial learning trajectory through iterative cycles of design, clinical interviews, and small-scale teaching experiments. Based on analysis of student behaviors during these empirical investigations, we revise the initial learning trajectory to a scale of proficiency levels to represent the overall progression of student reasoning.

Different task settings and different item parameter values produce different levels of challenge for students, especially before students have formed broader generalizations that apply across all instances of a problem type. Therefore, we further articulate the learning trajectory proficiency scale by delineating various classes of tasks relevant to the learning trajectory to create a two-dimensional learning trajectory *matrix*. This matrix representation reinforces the notion that progress through a learning trajectory is not strictly linear. It also serves as a framework to guide subsequent assessment task design and cataloguing of behavioral exemplars. In constructing a matrix of proficiency levels on one axis, with task classes on the other axis, we can create explicit conjectures about how students can accomplish movement through this conceptual space, and then examine the viability of these conjectures using various combinations of curricular design, classroom observation, and/or assessment design and psychometric modeling.

Based on the empirical data and theoretical analysis to this point, we also predict the possible responses to challenges or items that pertain to each proficiency level, across the task classes, and order them according to our observations and conjectures of their cognitive complexity. The result, a hypothesized *outcome space*, provides a further detailed framework with which to develop assessment items to probe in detail the learning trajectory's cognitive progression.

Exemplars of Student Reasoning

The semistructured, open-ended clinical interviews are revised as we learn more about the children's approaches to tasks. They are all recorded as video data. From the clinical interviews and teaching experiments, we identify behavioral exemplars of students' task responses to assist in designing appropriate items for piloting and field testing.

Item Design and Piloting

In developing the first set of assessment items, tasks that comprise structured combinations of problem type (construct, recognize, justify), response format (multiple choice, true/false, yes/no, and open-response), and task situation (physical materials and contexts referred to in the items) are created. Many are based directly on variations of clinical interview tasks, in order to attempt to elicit behaviors and responses that are as close as possible in assessment to the reasoning evoked by the clinical interviews.

We create assessment items for each relevant cell of the learning trajectory matrix (not all cells of the matrix will be assigned items, because some task classes are not relevant to particular proficiency levels). Assessment items in the various formats are piloted to elicit responses from small numbers of students of various ages, to see if the items are readily understood by the students in the medium of delivery, and to examine the range and content of the elicited responses.

Learning trajectories explicitly encompass a progression of conceptual understanding, from less to more sophisticated or complex, across time periods that may be as brief as a few weeks or as long as several years. Our research program assumes broad diversity of student experience with individual mathematical constructs and proficiency levels across different grade spans, particular with the equipartitioning construct, which is not yet taught as such in most schools; we attempt, therefore, not to presuppose particular matching of proficiency levels to grade levels. We assume, for instance, that some younger students may be much more adept with (higher) levels of a learning trajectory than might be predicted based on any particular curriculum. It is therefore critical to provide young children with items that allow them to express higher proficiency level mathematical reasoning but that are nonetheless appropriate for their literacy level. Likewise, older students with more advanced literacy may nonetheless not have experienced the lower proficiency levels of the equipartitioning learning trajectory. Therefore, a goal in our item design is to accommodate such diversity of language skills, ages, cognitive load, and mathematical proficiency levels. A challenge throughout this work is to build items that allow students to make their reasoning visible even if they are not adept with paper-and-pencil explanations. A recent phase of our research program is designing technologically enabled item generation environments that mimic the manipulatives used in the clinical interviews (Confrey & Maloney, in press).

Think-aloud interviews (Desimone & Le Floch, 2004; Wilson, 2005) are also conducted with students, to help capture more fully the range of responses elicited by some items as students solve them. Insights from the piloting and think-aloud interviews are incorporated, when appropriate, in revisions of items, proficiency level outcome spaces, ordering of the proficiency levels, the definitions of the proficiency levels, and delineations of distinct classes of tasks.

Item Field Testing

Field tests of the items are intended to accomplish several purposes that require large numbers of responses: (1) to evaluate the extent to which each item both is comprehensible to students and elicits responses that match the intent of the item's design; (2) to subject the entire bank of responses to statistical analysis in order to gain insight about the extent to which item responses and their inferred difficulty parallel the proficiency levels ordering; and finally, (3) to interpret the extent to which the combination of assessment items and learning trajectory are valid means of characterizing the progression of student reasoning about a construct. Considerations that contribute to field-test design include administration of individual items across broad enough grade bands to understand the range of grades across which particular proficiency levels are instructionally important, inclusion of some identical items in different test forms in a single grade band (linking items) and across grade bands (anchor items) to enable statistical calibration across all the item responses, and, finally, simply a sufficient number or responses for each item to ensure statistical validity.

Rubrics and Scoring

From the field-test results, we employ a two-stage process to build scoring rubrics. First, small subgroups of our research group (subject matter experts) examine all responses to an item, develop a draft scoring rubric based on those responses, and identify student response exemplars. Second, the entire research team evaluates the draft rubric for alignment of student responses and exemplars to the proficiency level definitions and the outcome space. Numerical values are used to represent the correspondence of classes of student responses to levels of the outcome space; some values are used to signal nonresponses as well as responses that were unintelligible or insufficient for assignment of score values. The rubrics also include qualitative codes that represent particular strategies and misconceptions that could be useful as diagnostic information for teachers.

The field-test items are then submitted to a team of trained third-party test scorers, who use the rubrics to independently generate scores for all items and all respondents. This reduces bias in the scoring and generates the scoring data to which measurement models will then be fit. In addition, the two-stage rubric development, along with third-party scoring, provide one set of data on which subsequent item revisions can be based.

Analysis and Interpretation of Field-Test Data and Validity of Learning Trajectory

Field-test score data are initially subjected to item response theory (IRT) modeling. IRT analysis of assessment item scores facilitates evaluation of relative item difficulty in relation to the learning trajectory matrix. It provides an initial basis for reviewing and revising particular items (especially if there are many incoherent or unintelligible student responses, or if the apparent item difficulty differs greatly from that of presumably similar items) and, overall, for further development of validity arguments for the learning trajectory.

Iterative Nature of the Methodology

The creation of learning trajectories and diagnostic assessments by the DELTA group is deeply grounded in research on children's learning. The genesis of children's understanding of the mathematical content and reasoning does not mirror a logical-sequential analysis of the discipline of mathematics (Confrey, 2012; Clements & Sarama, Chapter 1, this volume). However, a goal of mathematics education is for students to understand the structure of mathematics as a discipline. In particular, domain goals of learning trajectories include successive levels of generalization about mathematical constructs and reasoning.

At multiple points in this process, revisions to one or more of the components of the project are anticipated, based on newly generated data. We carefully review all new data (clinical interview, think-aloud, assessment items, field-test data, etc.) with respect to representations of the learning trajectory. We assume that adjustments may be needed to the learning trajectory matrix. The research program is open to generating new hypotheses and will conduct additional smaller-scale studies on parts of the learning trajectory to attempt to clarify relationships, definitions, and/or ordering among one or more proficiency levels and/or their corresponding assessment items, and reexamining the design of upcoming phases of the program in light of these iterations. Such cycles of design-implementation-analysis-revision are crucial to generating robust learning trajectories and diagnostic assessments that are valid for classroom use. It is important to note that this kind of research and development program is necessarily empirical, theoretically grounded, highly iterative, and time- and resource-intensive.

Ellie's strategy is one known as *mark-all* (Lamon, 1996). She recognized that for multiple wholes, even though her strategy for making a set of fair shares involves splitting each whole into the same number of parts as people, the final share per person is nonetheless represented by the unit fraction $1/n$ (n = the number of persons) of the original collection of pizzas. But her generalization also recognizes two equivalent ways of answering, depending on the unit to which she refers (i.e., the "referent unit"): Each person's share is *one third of the original total* amount of pizza, as well as *two thirds of a single* pizza.

REPRESENTING PROFICIENCY LEVELS AND TASK CLASSES

Delineating the variety of cognitive reasoning elements in our framework, combined with the three cases, led us to propose the current form of our learning trajectory. The learning trajectory for equipartitioning is proving to be foundational for the development of the whole rational number reasoning space under construction. We argue that it is foundational because through our research we are demonstrating how the roots of division and multiplication sit within the trajectory, with key links to geometric ideas of length, area, similarity, and scaling. Further, we are demonstrating in related work the key linkages between this work and the development of fractions, ratio relations, and the construct of *a/b*-as-operator. We claim that these three key ideas form the primary structures necessary for a robust understanding of rational number reasoning (Confrey, Maloney, Nguyen et al., 2008; Confrey et al., 2009).

A vertical display of learning trajectory proficiency levels was developed to incorporate the three cases of equipartitioning, as well as precise statements of the knowledge and skills that we conjectured should accrue to students through successfully solving case-specific challenges. Table 3.1 delineates the 16 proficiency levels[3] of the equipartitioning learning trajectory, listed from less sophisticated at the bottom to more sophisticated at the top.

For each proficiency level the outcome space, an ordered list (also ordered from least to most sophisticated) combined student reasoning results we had observed in the clinical interviews and predictions of those we expected as we expanded the tasks into assessment items. The outcome spaces are designed (again, iteratively) to convey sufficient detail to understand the cognitive behaviors associated with the proficiency level. They served to guide the development of paper-and-pencil assessment items.

To ensure consideration of student understanding across the range of equipartitioning cases and across the K through 7 grade levels, a two-dimensional display of proficiency levels and task types was added. The proficiency levels in one dimension were arrayed with a second dimension known as *task classes* (Table 3.2).

TABLE 3.1. Proficiency Levels for the Equipartitioning Learning Trajectory

No.	Proficiency Level
16	Generalize that *a* objects shared among *b* persons results in *a*/*b* objects per person, applying strategies based on both the distributive property and ratio reasoning, and asserting their equivalence.
15	Apply distributive property to multiple wholes, demonstrating equivalence of multiple ways of equipartitioning over breaking or fracturing
14	Make factor or split-based changes in the number of objects, the number of people sharing, and/or the size of fair shares, and predict the effects on the other variables (co-splitting)
13	Predict the outcome of a composition of splits on multiple wholes
12	Equipartition multiple wholes among multiple persons and name the resulting shares in relation to referent units
11	Assert that a whole can be equipartitioned for all natural numbers greater than 1 (continuity principle)
10	Demonstrate equivalence of same splits resulting in non-congruent parts across or within methods of non-prime equipartitioning (property of equality of equipartitioning)
9	Demonstrate and justify how extra shares can be redistributed for fewer people (additive changes) sharing collections (equipartitioning over breaking to quantify compensation).
8	Demonstrate and justify the effect of factor-based changes in the number of persons sharing on the size of a share, and vice versa, for collections or single wholes (inverse variation to quantify compensation)
7	Predict, demonstrate, and justify outcomes, including with multiple methods, of compositions of splits on collections or a single whole.
6	Predict (qualitatively) the effect of changes in the number of people sharing on the size of shares (qualitative compensation).
5	Re-assemble equal groups or parts to produce the collection or the single whole as "*n* times as many" or "*n* times as much" as a single group or part
4	Name the shares resulting from equipartitioning collections or single wholes, in relation to referent units
3	Justify results by counting, stacking, arrays, or patterns
2	Equipartition a single whole (circles and rectangles)
1	Equipartition collections

TABLE 3.2. Task Classes for Equipartitioning Learning Trajectory

A	B	C	D	E	F	G	H	I	J	K	L	M
Case A				Cases B and C						Case C		
Collections	2-split (Rect/Circle)	2^n split (Rect)	2^n split (Circle)	Even split (Rect)	Odd split (Rect)	Even split (Circle)	Odd split (Circle)	Arbitrary integer split	$p = n + 1$; $p = n - 1$	p is odd, and $n = 2i$	$p >> n$, p close to n	all p, all n (integers)

TABLE 3.3. Equipartitioning Learning Trajectory Matrix (Grades K–8)

		A	B	C	D	E	F	G	H	I	J	K	L	M
	Task Classes →	Collections	2-split (Rect/Circle)	2^n split (Rect)	2^n split (Circle)	Even split (Rect)	Odd split (Rect)	Even split (Circle)	Odd split (Circle)	Arbitrary integer split	$p = n + 1$; $p = n\text{-}1$	p is odd, and $n = 2^i$	$p >> n$, p close to n	all p, all n (integers)
Proficiency Levels														
16	*Generalize*: a among $b = a/b$													
15	*Distributive property*, multiple wholes													
14	*Co-splitting*													
13	*Compositions* of splits, multiple wholes													
12	Equipartition *multiple wholes*													
11	Assert *Continuity principle*													
10	*Property of Equality of Equipartitioning*													
9	Redistribution of shares (*quantitative*)													
8	Factor-based changes (*quantitative*)													
7	*Compositions* of splits; *factor-pairs*													
6	*Qualitative compensation*													
5	*Re-assemble*: n times as much													
4	*Name* a share w.r.t. the referent unit													
3	*Justify* the results of equipartitioning													
2	*Equipartition single wholes*													
1	*Equipartition Collections*													

Note. Proficiency levels form the vertical dimension, listed in the left column of the table. Task classes, listed along the top row, form the horizontal dimension.

Task classes were ordered to accommodate observations that different examples of problems are easier or harder for students. For example, 2-splits and powers of 2-splits (4, 8, and 16) are easier than the other even splits (6, 10, 12). Even splits are easier for most children than odd splits, especially when the task targets compositions of splits. For instance, unless students have previously been introduced to the *Y* shape of the 3-split on a circle, they are more likely to be successful with a 6-split (applying a 2-split to the circle to begin with, then 3-splitting each resulting semicircle) than with a 3-split.

After students begin to work with equipartitioning of collections and single wholes[4] across a variety of split values, they are presented with particular cases of multiple wholes, which vary in difficulty for children. For example, sharing four wholes among three people is easier, for most children of similar grades in school, than sharing five wholes among three people. Students will commonly deal out a single whole to each person (Case A behavior) and then split the remaining wholes (Case B behavior). Empirically, it proved easier for students to split a single whole among three people that to split two wholes.

The entire learning trajectory matrix is displayed in Table 3.3.

ITEM CREATION, FIELD TESTING, AND RUBRIC DEVELOPMENT

The next step in the process was to build, pilot, and field-test items for each cell of the matrix for which the combination of the row and column can result in a relevant task situation. A pool of approximately 120 separate items was developed to span relevant combinations of proficiency levels and task classes of the learning trajectory matrix. In the field test of equipartitioning learning trajectory items, we administered 116 items, distributed across the 16 proficiency levels and 13 task classes. The items were administered to 4,800 students ranging from kindergarten to grade 7 in seven elementary and three middle schools in four North Carolina school districts. Of these districts, one was urban, one was rural, and two were mixed suburban/rural. To make the administration of this relatively large item-pool practicable, we used a complex booklet design with 33 booklets/test forms (see Frey, Hartig, & Rupp, 2009; Rutkowski, Gonzalez, Joncas, & von Davier, 2010). Specifically, 10 forms consisted of six items each for grades K–1 and 2–3, 8 forms consisted of eight items each for grades 4–5, and 5 forms consisted of eight items each for grades 6–7. The forms were designed for completion within a single class period to minimally disrupt teachers' normal instructional schedules.

Outcome spaces were developed for each of the items. All responses were hand-scored as described above. We provide here an example of one of the items (Figure 3.11). This item corresponds to Level 10, the property of equality of equipartitioning. Its outcome space is described in Table 3.4

Ann and Beth have brownies that are the same size. The children cut their brownies as shown in the drawings below. A piece of Ann's brownie is shaded in on the left. A piece of Beth's brownie is shaded on the right.

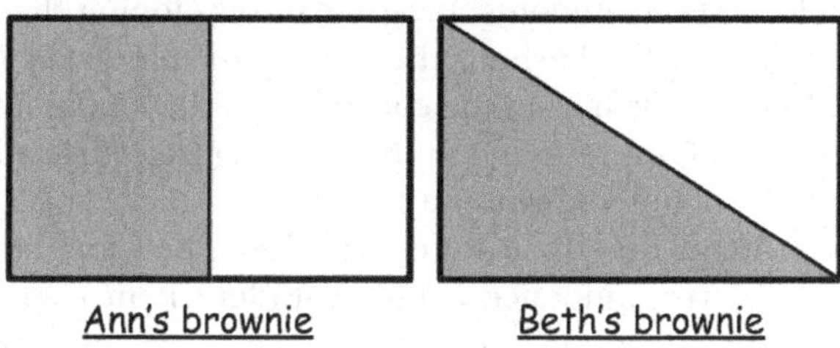

Is the piece of Ann's brownie bigger, smaller, or the same size as the piece of Beth's brownie?
Circle one:
Smaller Bigger the Same

FIGURE 3.11. A sample assessment item for proficiency level 10

and corresponds to the display of the proficiency levels, with the least sophisticated responses at the bottom and the most sophisticated at the top. Rubrics were developed based on the outcome spaces, then adjusted for the range of student responses actually received. All items were then scored by independent third-party team (supported by Pearson).

Role of IRT Analysis in Validation of Assessment Items and Learning Trajectories

Evidence-based assessment relies on inferring students' cognition from observed student responses to assessment items. Making such inferences re-

TABLE 3.4. Levels of Outcome Space for Proficiency Level 10 of the Equipartitioning Learning Trajectory

- Argues for equivalence of non-congruent equipartitions formed based on the basis of same types of splits of congruent wholes or equal-size parts.
- Argues for equivalence of non-congruent equipartitions formed based on composition (combining areas) and decomposition (separating areas).
- Argues for equivalence of non-congruent equipartitions formed based on qualitative compensation.
- Argues for equivalence of non-congruent equipartitions formed based on perception or assertion.
- Does not recognize equivalence of non-congruent parts.

lies on interpreting the item responses in relation to an underlying model of cognitive development—in this case, the learning trajectory. Eventually, for diagnostic assessment purposes, students' responses to any particular suite of items (an assessment) must not only correspond to the underlying learning trajectory but must also provide the basis of interpretable (valid) feedback to teachers about students' progress in developing their reasoning about the construct. Establishing the validity of inferences from student responses depends further on the construct validity of the underlying learning trajectory itself, represented in the learning trajectory matrix and further specified through item response studies.

The assessment items described here were developed for the purpose of interpreting the correspondence between student item responses and the learning trajectory (the underlying model of cognitive progression with respect to the equipartitioning construct) and to form a foundation for an assessment tool for diagnostic feedback about students' progress in developing their reasoning about the equipartitioning construct.

We undertook IRT-based analysis of the scored student responses, to examine and interpret the correspondence between the equiparitioning proficiency levels and the difficulty of the various assessment items, and to provide a further, detailed empirical basis for revising and stabilizing the items and learning trajectory representation in relation to each other. Rasch explanatory IRT models (deBoeck & Wilson, 2004) were used to analyze the scored student response data, a model chosen due to the relatively small number of responses to each item. A technical description of this analysis is detailed by Confrey, Rupp, Maloney, and Nguyen (in review). We comment here on results from this analysis that are most relevant to this chapter.

One validation step is the comparison of the item proficiency-level classification to the item difficulty estimates. These results for mean item difficulty and for the distribution of item difficulty are displayed in Figures 3.12a and 3.12b, respectively. The general linear trend of the item difficulties indicates that on average, items designed to measure more complex equipartitioning proficiencies have higher item difficulty, and vice versa. However, there is a great deal of variability in the item difficulties at any given proficiency level.

There is less empirical support for a generally linear increase in item difficulty across task classes, as shown in Figure 3.13; the mean item difficulty appears in Figure 13a, and distribution of those values appear in Figure 13b. The difficulty of items in task classes I through M, taken together, appear to represent a cluster of items of higher mean difficulty relative to task classes A through H.

A Wright map display (not shown), calibrating student performance with item difficulty, provided a great deal of information about item difficulty ordering across student grade band and empirical support for the

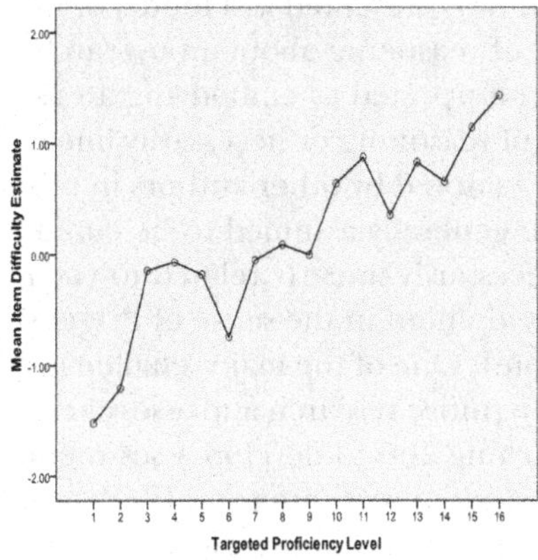

FIGURE 3.12a. Mean item difficulties of assessment items classified to different proficiency levels

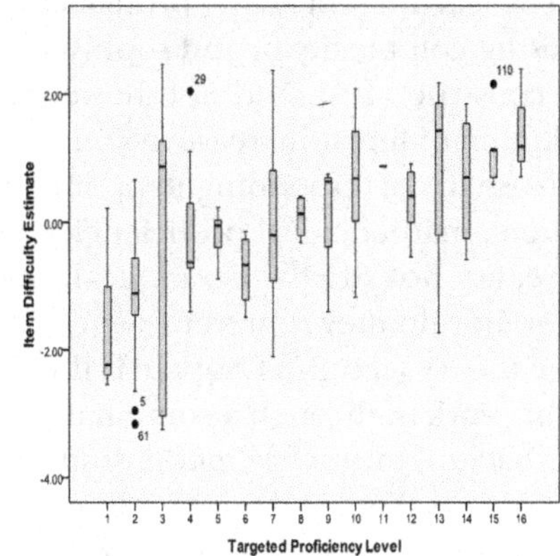

FIGURE 3.12b. Item difficulty distributions of items classified to different proficiency levels

general ordering of items in relation to the framework for understanding. Combined with item fit information for different grade bands, the Wright map called our attention to previously unpredicted issues in the overall design of particular types of items in Case B (emerging relations and properties), including the ordering and definition of several proficiency levels, and laid the groundwork for additional small-scale investigations of those portions of the learning trajectory (Yilmaz, 2012).

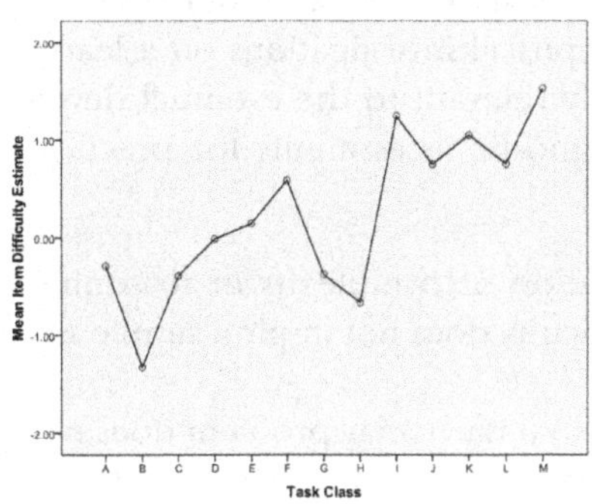

FIGURE 3.13a. Mean item difficulties of assessment items in different task classes

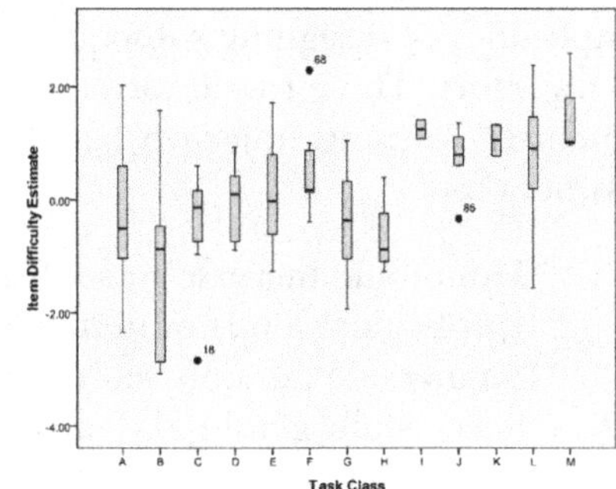

FIGURE 3.13b. Item difficulty distributions for assessment items in different task classes

The learning trajectory proficiency levels are presented as a model of increasing complexity in and sophistication of reasoning about an overarching construct. It should not, however, be interpreted as embodying an assumption of linear increase in complexity of reasoning or necessarily linear increase in corresponding item difficulty. As noted by other authors in this volume, mathematical proficiencies, while generally assumed to be cumulative, are not strictly hierarchical, nor necessarily linearly related to each other, nor do they represent stages of development in the sense of Piaget's stage theory (see also Chapter 1, this volume). One of the major challenges of this work is, through a combination of cognitive research and assessment approaches, to flexibly model student learning and to develop assessment approaches that (1) do not underestimate student reasoning and (2) detect and support students' reorganization of their understanding to engage in more complex tasks and generate and apply increasingly robust generalizations in ever more resourceful ways.

Our working hypothesis regarding the relationship between assessment items and their corresponding proficiency levels is as follows: (1) a learning trajectory embodies an approximately monotonically increasing sophistication of student reasoning, expressed as proficiency levels; and (2) the pool of items corresponding to the learning trajectory's proficiency levels should exhibit a corresponding monotonic pattern of difficulty. IRT's assumption of unidimensionality of a construct constitutes a reasonable proxy for monotonicity of item difficulty. IRT analysis therefore provides a quantitative means to facilitate interpretation of the correspondence of proficiency levels with item difficulty. Several caveats, however, challenge straightforward interpretations of Rasch analysis, technically complicate the design of assessment items and field test for statistical analysis, and highlight the complexities of designing subsequent empirical investigations on a learning trajectory. These caveats are especially relevant to the eventual development of learning trajectory-based diagnostic assessments for practical classroom use:

1. Monotonic increase in sophistication or complexity of reasoning (proficiency level) or in item difficulty does not imply a simple *linear* increase on some score scale.
2. Successfully getting the "answer" to a particular problem does not necessarily mean that a student (a) understands or can make an argument that supports the answer, (b) can reason flexibly in different problem contexts, or (c) can solve a problem with the same numerical solution when posed in a different way. These considerations make more difficult the job of recognizing gaps in students' reasoning capabilities and embedding these in assessments based on learning trajectory approaches.

3. The second dimension of the learning trajectory matrix (task classes) may interact in complex ways with the proficiency levels, complicating or disproving an assumption of monotonic increase in item difficulty with proficiency levels or task classes, as well as the long-term goal of providing rapid diagnostic feedback to teachers. For example, specific item parameter values may make items from lower proficiency levels disproportionately more difficult than researchers or teachers might expect. For example, we anticipate that for many students, items in task class H, I, or J from proficiency level n could prove more difficult than items in task class D or E from proficiency level $n + 1$ or $n + 2$, (i.e., higher proficiency levels).
4. Heterogeneity in student background or prior knowledge (due to cultural, family, or educational background, and including the curriculum used or students' numerical proficiency) could locally complicate any hypothesis of monotonicity of item difficulty or proficiency levels for one or more groups within a sampled student population. For instance, students who have been previously introduced to fractions might readily provide numeric solutions to items that have fractional answers, but they may demonstrate weaknesses in the mathematical practices (naming of results, or justifying their reasoning) relative to other students.
5. One of the goals of learning trajectory research is to describe the proficiency and progression of student reasoning about mathematical constructs across a broad range of student ages: Diagnostic tools should support teachers in adjusting instruction for subgroups of students, for those whose mastery of big ideas is lagging behind that of their classmates, and accelerate their progression to grade level, as well as to recognize and support students who are progressing well with respect to grade level expectations. Differences in student age, reading ability, and English language proficiency present ongoing challenges to the design of items for both age and linguistic appropriateness and mathematical and cognitive interpretability for a single construct across a broad age range.
6. For diagnostic purposes in formative settings, unidimensionality of construct, or demonstration of linearity of some scoring regimen for associated items, may not be a useful criterion of validity of a learning trajectory or a related assessment tool.

In sum, we do not anticipate a simple, hard-and-fast conclusion to questions regarding the validity of learning trajectories or (eventually) of diagnostic assessments based on learning trajectories. We point to the perspective of Kane (2006) for a more nuanced view that validity of an assessment depends heavily on the context and consequences of the use of that assessment, and on the value, to teacher and student, of the feedback resulting from the assessment. This is true especially in a low-stakes situation such as in-class formative uses of diagnostic assessments and depends on interpretation of the results in relation to the use to which the assessment is put. This observation may relax the stringency of some statistical assumptions in modeling the students' performances and suggests that multiple statistical (psychometric) models should be employed to provide additional insight into the patterns of student cognitive progression in relation to constructs based on learning trajectories.

IRT analysis is one tool with which to investigate the relative difficulty of items and the correspondence between the learning trajectory levels and the items. Subsequent statistical research will include the use of other models that do not rely so heavily on assumptions of unidemensionality, such as diagnostic classification models, also sometimes referred to as cognitive diagnostic models (Rupp & Templin, 2008; Rupp, Templin, & Henson, 2010), and Bayesian networks (Almond, Mislevy, Steinberg, Williamson, & Yan, in press). Designing for use of these models will likely necessitate additional new items and further piloting and field testing.

CONCLUSIONS

This chapter outlines the general argument for a learning trajectory approach to the development of big ideas over time in mathematics. In addition, it describes the methodology used to develop the first learning trajectory in our project on the development of rational number reasoning. We emphasize the importance of a thorough review of literature and acts of synthesis necessary to use that review to drive subsequent research forward. Clinical interviews and short teaching experiments are necessary to test, refine, and complete the proposed learning trajectory. Because this does not involve longitudinal study, but rather cross-sectional analysis, subsequent further validation would need to be undertaken. However, based on the proposed learning trajectory, another source of validation is the design of assessment items and the conduct of large-scale assessment. We reported on how IRT analysis was used to show that the assessment items corresponding to higher levels of the trajectory tended to be more difficult. Overall, all of our students showed difficulty with the equipartitioning items, suggesting that this is an area that merits more curricular attention in the future.

The final phase of this component of the research involves successive co-refinements of the learning trajectory and the items, with the intention that the learning trajectory and the items/responses are mutually coherent, permit interpretations of student progress and obstacles with respect to the learning trajectory, and for which the learning trajectory provides instructionally valid basis on which a teacher may adjust instruction. Iterative co-refinements of learning trajectory and assessment items lead, we postulate, to stabilization of both. This can be the basis of several efforts: (1) diagnostic assessments and learning progress profiles for big ideas in mathematics, as a source of (formative) feedback to teachers and students about students' progress in developing proficiency in mathematical ideas; (2) a tool and representation that teachers can use to improve their own understanding of the most likely progressions in students' reasoning and to strengthen their own instructional planning for and responses to student mathematical reasoning and behaviors; (3) a basis for preservice and in-service content professional development; and (4) a framework for curricular design.

AUTHOR'S NOTE

We acknowledge and appreciate funding support from the National Science Foundation (Grants DRL 758151 and DRL 0733272) and Qualcomm. The views and opinions expressed in this chapter represent the thinking of the authors and not necessarily those of the funders.

NOTES

[1]In the research synthesis, we originally identified two cases of multiple whole, Case C, in which there are more people than objects, and resulting in proper fractions, and Case D, in which there are more objects than people, and resulting in improper fractions or mixed numbers. However, in clinical interviews, the strategies that students used to solve these situations depended on which "case" students had been exposed to first, so we combined the two cases into one, now referred to as Case C (multiple wholes).

[2]Pothier and Swada (1983) called this construction of shares a *composition of factors*. We rename it *composition of splits* to emphasize that it shows how splitting can anticipate the idea of factors.

[3]While the movement through the methodology is described as a linear sequence, it entails multiple iterations. The proficiency levels of the learning trajectory were originally outlined from the synthesis work, then revised and reorganized in light of the analysis of the clinical interviews and the goal of item construction. They were revised again after piloting and field testing as we identified the various responses from children. It was revised further in the wake of field-test data analysis and subsequent investigations into several proficiency levels related to naming and justifying (levels 3–5), emergent relations for single wholes, and co-splitting (level 15).

[4]When the term *collection* is used, cases are limited to totals which can be evenly divided (creating whole number outcomes) among the persons sharing; when the term *single whole* is used, we constrain our examples to circles and rectangles; and when the term *multiple wholes* is used, the tasks typically do not result in shares that have only whole shapes.

REFERENCES

Almond, R. G., Mislevy, R. J., Steinberg, L. S., Williamson, D. M., & Yan, D. (in press). *Bayesian networks in educational assessment.* New York, NY: Springer.

Battista, M. T. (2009, August). *Cognition Based Assessment.* Paper presented at the Realizing the Potential for Learning Trajectories Research to Serve as Evidence and Validation for Standards and Related Assessments Conference, Raleigh, NC.

Behr, M. J., Cramer, K., Harel, G., Lesh, R., & Post, T. (2011). *The rational number project.* Retrieved from http://www.cehd.umn.edu/rationalnumberproject/

Charles, K., & Nason, R. (2000). Young children's partitioning strategies. *Educational Studies in Mathematics, 43*(2), 191–221.

Clements, D. H., & Sarama, J. (2004). Learning trajectories in mathematics education. *Mathematical Thinking and Learning, 6*(2), 81–89.

Clements, D. H., Wilson, D. C., & Sarama, J. (2004). Young children's composition of geometric figures: A learning trajectory. *Mathematical Thinking and Learning, 6*(2), 163–184.

Confrey, J. (1988). Multiplication and splitting: Their role in understanding exponential functions. In M. Behr, C. LaCompagne, & M. Wheeler (Eds.), *Proceedings of the Tenth Annual Meeting of the North American Chapter of the International Group for the Psychology of Mathematics Education* (pp. 250–259). DeKalb, IL: Northern Illinois University Press.

Confrey, J. (1994). Splitting, similarity, and rate of change: a new approach to multiplication and exponential functions. In G. Harel & J. Confrey (Eds.), *The development of multiplicative reasoning in the learning of mathematics* (pp. 293–332). Albany, NY: State University of New York Press.

Confrey, J. (2006). The evolution of design studies as methodology. In R. K. Sawyer (Ed.), *The Cambridge handbook of the learning sciences* (pp. 135–152). New York, NY: Cambridge University Press.

Confrey, J. (2008, July). *A synthesis of the research on rational number reasoning: A learning progressions approach to synthesis.* Paper presented at the 11th International Congress of Mathematics Instruction, Monterrey, Mexico.

Confrey, J. (2012). Better measurement of higher-cognitive processes through learning trajectories and diagnostic assessments in mathematics: The challenge in adolescence. In V. Reyna, S. Chapman, M. Dougherty, & J. Confrey (Eds.), *The adolescent brain: Learning, reasoning, and decision making* (pp. 155–182). Washington, DC: American Psychological Association.

Confrey, J., & Maloney, A. P. (2012). Next generation digital classroom assessment based on learning trajectories in mathematics. In C. Dede & J. Richards (Eds.), *Steps toward a digital teaching platform* (pp. 134–152). New York, NY: Teachers College Press.

Confrey, J., Maloney, A., & Nguyen, K. H. (2008). Rational number reasoning database. Retrieved from http://gismo.fi.ncsu.edu/index.php?module=Search

Confrey, J., Maloney, A., Nguyen, K., Mojica, G., & Myers, M. (2009). Equipartitioning/splitting as a foundation of rational number reasoning using learning trajectories. In M. Tzekaki, M. Kaldrimidou, & H. Sakonidis (Eds.), *Proceedings of the 33rd Conference of the International Group for the Psychology of Mathematics*

Education (Vol. 2, pp. 345–352). Thessaloniki, Greece: International Group for the Psychology of Mathematics Education.

Confrey, J., Maloney, A., Nguyen, K., Wilson, P. H., & Mojica, G. (2008, April). *Synthesizing research on rational number reasoning.* Paper presented at the Research Pre-Session of the Annual Meeting of the National Council of Teachers of Mathematics. Salt Lake City, UT.

Confrey, J., Maloney, A. P., Wilson, H., & Nguyen, K. (2010, May). *Understanding over time: The cognitive underpinnings of learning trajectories.* Paper presented at the Annual Meeting of the American Educational Research Association, Denver, CO.

Confrey, J., Rupp, A. A., Maloney, A. P., & Nguyen, K. (unpublished). Toward developing and representing a novel learning trajectory for equipartitioning: Literature review, diagnostic assessment design, and explanatory item response.

Confrey, J., & Scarano, G. H. (1995, October). *Splitting reexamined: Results from a three-year longitudinal study of children in grades three to five.* Paper presented at the Annual Meeting of the North American Chapter of the International Group for the Psychology of Mathematics Education, Columbus, OH.

Cooper, H. (1998). *Synthesizing research.* Thousand Oaks, CA: Sage.

deBoeck, P., & Wilson, M. (Eds.). (2004). *Explanatory item response models: A generalized linear and nonlinear approach.* New York, NY: Springer.

Desimone, L. M., & Le Floch, K. C. (2004). Are we asking the right questions? Using cognitive interviews to improve surveys in education research. *Educational Evaluation and Policy Analysis, 26*(1), 1–22.

Elmore, R. F. (2002). *Bridging the gap between standards and achievement.* Washington, DC: Albert Shanker Institute.

Empson, S. B., & Turner, E. (2006). The emergence of multiplicative thinking in children's solutions to paper-folding tasks. *Journal of Mathematical Behavior, 25,* 46–56.

Frey, A., Hartig, J., & Rupp, A. A. (2009). An NCME instructional module on booklet designs in large-scale assessments of student achievement: Theory and practice. *Educational Measurement: Issues and Practice, 28*(3), 39–53.

Ginsburg, H. P., Kossan, N. E., Schwartz, R., & Swanson, D. (1983). Protocol methods in research on mathematical thinking. In H. P. Ginsburg (Ed.), *The Development of Mathematical Thinking* (pp. 7–47). New York, NY: Academic Press.

Hunting, R. P., & Sharpley, C. F. (1991). Pre-fraction concepts of preschoolers. In R. P. Hunting & G. Davis (Eds.), *Early fraction learning* (pp. 9–26). New York, NY: Springer-Verlag.

Kane, M. T. (2006). Content-related validity evidence in test development. In M. S. Downing & M. T. Haladyna (Eds.), *Handbook of test development* (pp. 131–153). Mahwah, NJ: Lawrence Erlbaum Associates.

Lamon, S. J. (1996). The development of unitizing: Its role in children's partitioning strategies. *Journal for Research in Mathematics Education, 27*(2), 170–193.

Maloney, A. P., & Confrey, J. (2010, July). *The construction, refinement, and early validation of the equipartitioning learning trajectory.* Paper presented at the 9th International Conference of the Learning Sciences, Chicago, IL.

Mislevy, R. J., Steinberg, L. S., & Almond, R. A. (2003). On the structure of educational assessments. *Measurement: Interdisciplinary Research and Perspectives, 1*, 3–67.

Mislevy, R. J., Steinberg, L. S., Almond, R. A., & Lukas, J. F. (2006). Concepts, terminology, and basic models of evidence-centered design. In D. M. Williamson, I. I. Bejar, & R. J. Mislevy (Eds.), *Automated scoring of complex tasks in computer-based testing* (pp. 15–48). Mahwah, NJ: Erlbaum.

Mojica, G. (2010). *Preparing pre-service elementary teachers to teach mathematics with learning trajectories.* Unpublished doctoral dissertation, North Carolina State University, Raleigh, NC.

Opper, S. (1977). Piaget's clinical method. *Journal of children's Mathematical Behavior, 5*, 90–107.

Pellegrino, J. W., Chudowsky, N., & Glaser, R. (2001). *Knowing what students know: The science and design of educational assessment.* Washington, DC: National Academy Press.

Pepper, K. L. (1991). Preschoolers' knowledge of counting and sharing in discrete quantity settings. In R. Hunting & G. Davis (Eds.), *Early fraction learning* (pp. 103–129). Heidelberg, Germany: Springer-Verlag.

Piaget, J. (1970). *Genetic epistemology.* New York, NY: W.W. Norton.

Pothier, Y., & Sawada, D. (1983). Partitioning: The emergence of rational number ideas in young children. *Journal for Research in Mathematics Education, 14*(4), 307–317.

Rupp, A. A., & Templin, J. L. (2008). Unique characteristics of diagnostic classification models: A comprehensive review of the current state-of-the-art. *Measurement, 6*, 219–262.

Rupp, A. A., Templin, J. L., & Henson, R. (2010). *Diagnostic measurement: Theory, methods, and applications.* New York, NY: Guilford Press.

Rutkowski, L., Gonzalez, E., Joncas, M., & von Davier, M. (2010). International large-scale assessment data: Issues in secondary analyses and reporting. *Educational Researcher, 39*, 142–151.

Simon, M. A. (1995). Reconstructing mathematics pedagogy from a constructivist perspective. *Journal for Research in Mathematics Education, 26*(2), 114–145.

Toluk, Z., & Middleton, J. A. (2003). The development of children's understanding of the quotient: A teaching experiment. Arizona State University: *Unpublished.*

Wilson, M. (2005). *Constructing measures: An item response modeling approach.* Mahwah, NJ: Lawrence Erlbaum Associates.

Wilson, P. H. (2009). *Teachers' uses of a learning trajectory for equipartitioning.* Unpublished doctoral thesis, North Carolina State University, Raleigh, NC.

Yilmaz, Z. 2011. *Toward understanding of students' strategies on reallocation and co-variation items: In relation to an equipartitioning learning trajectory.* Unpublished master's thesis, North Carolina State University, Raleigh, NC.

CHAPTER 4

TWO APPROACHES TO DESCRIBING THE DEVELOPMENT OF STUDENTS' REASONING ABOUT LENGTH

A Case Study for Coordinating Related Trajectories

Jeffrey E. Barrett and Michael T. Battista

ABSTRACT

This chapter compares two different learning trajectories describing length and its measurement. By analyzing the origin of these two trajectories within the respective research projects, we offer a case study for the coordination and consolidation of different trajectories. The analysis commences with a detailed comparison of both learning trajectories.

Comparative analysis of these research projects and the resulting accounts of student learning and development in the form of learning trajectories helped

Learning Over Time: Learning Trajectories in Mathematics Education, pages 97–124.

us pursue several questions about the use of trajectories as integrative tools for research, curriculum development, evaluation, and standard setting. This case study illustrates the importance of finding broadly common terminology for describing the growth of children's knowledge and strategies for a topic—length measurement in this case. We identified a common partial sequence of increasingly sophisticated reasoning consistent with both trajectories but did not find a completely consistent system allowing a comprehensive coordination. Nevertheless, by relying on this common partial sequence, we examined the correspondence of sections of both research-based trajectories and produced a visual mapping to represent our analysis. We developed a rubric of five types of correspondence between related trajectories and used this rubric to characterize the degree of fit between these two trajectories. Thus, we provide one example that blends related trajectories to forge a common recommendation for content-specific pedagogy that may help teachers improve instruction.

In this chapter, we compare how two research projects investigate and describe the development of elementary students' reasoning about length, and conduct a detailed comparison of the learning trajectories suggested by these research projects. We also address the question, "What can a comparative analysis across research projects tell us about patterns of learning, effects of instruction, and the construct of learning trajectories?" Comparative analysis of these different research projects has helped us identify, clarify, and pursue several questions about the potential of trajectories as integrative tools for research, curriculum development, evaluation, and standard setting: Is it possible to establish common terminology that permits straightforward comparison of trajectories for the same topic? In what ways may we identify similarities in the learning trajectories written for researchers and learning trajectories written for teachers, curriculum developers, and organizations specifying state standards? Can we blend related trajectories to forge a common recommendation for content-specific pedagogy to help teachers improve instruction?

Battista's research project (2001, 2012) describes a set of levels of sophistication in students' concepts, reasoning, and strategies; he defines a level of sophistication as a qualitatively distinct type of cognition that occurs within a hierarchy of cognition levels for a specific domain.[1] These levels form plateaus in the cognitive terrain that students ascend in their individual routes from their beginning states of knowledge to the formal knowledge targeted by instruction (see Figure 4.1). Advances to the higher plateaus indicate increasingly sophisticated thinking as a student develops increasingly abstract and powerful patterns of reasoning about the concept of length and its measurement. Battista's set of levels describes actual student reasoning observed in grades 1–5—many fifth graders were at the lower levels. However, jumps between successive levels typically occur after several days of high quality instruction.

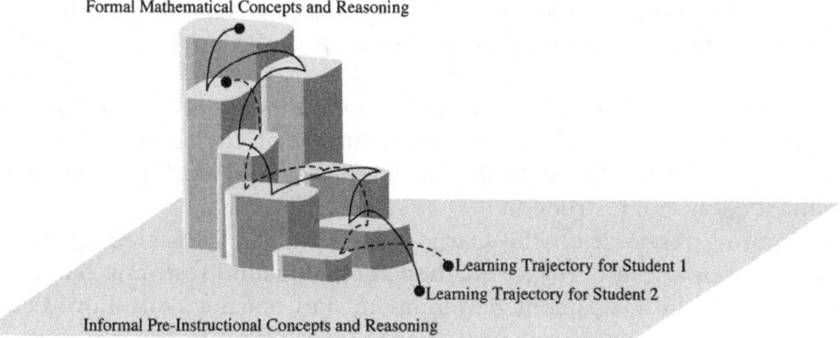

FIGURE 4.1. Battista's levels of sophistication plateaus

Researchers Clements, Sarama, and Barrett (2007; referred to as CSB) describe a learning trajectory by picturing a cumulative set of layers in a cross-sectional view of sedimentary material (as in sedimentary rock formations). The set of layers represents the accumulation of educational competence and the increasing depth of cognitive resources established over time (see Figure 4.2; also, see Figure 1.1 in Chapter 1, this volume). Advances to the upper layers indicate increasingly sophisticated thinking as a student

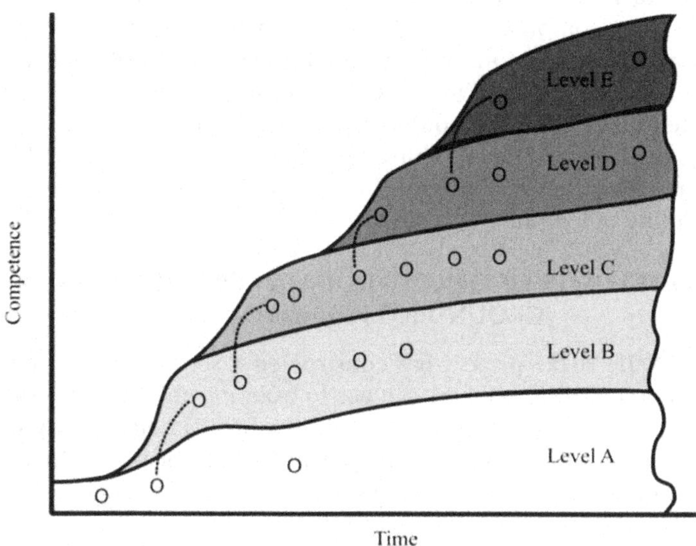

FIGURE 4.2. Clements, Sarama & Barrett's trajectory relating competence with time. *Note:* Clements, Sarama & Barrett's trajectory through layers relating competence with time: showing an insightful student making prolonged progress, within an excellent instructional setting, as part of a conceptually and broadly well-prepared population.

develops increasingly abstract patterns of thinking in response to tasks and challenges. The set of layers taken altogether provides a representation of the ordered developmental progress of children as they develop increasingly integrated knowledge and strategies for measurement of length. Each *O* in a given layer represents a specific incident that fits the layer. This figure represents an idealized account that would span several years' time, given a specific sequence of instruction.

Integrating research on the learning of particular core mathematical topics is important not only for mathematics education research, but also for standards and assessment development and teacher education. Our comparison of research projects that examine children's developing reasoning about length and length measurement is a case study in the comparison and extraction of commonalities of research findings. By comparing the research constructs, analyses, and findings of different research groups for the same concept, and elucidating differences and commonalities, we aim to illustrate the viability, generality, and challenges of developing and using learning trajectories. We also present a method of conceptualizing the comparison of two trajectories and examining the fit between them. This comparative analysis yields an essential account of the components of students' development of reasoning about length measurement. At the same time, we address several questions that frame our analysis.

This chapter comprises three parts. First, we describe the objectives, methods of data collection, and grounding frameworks of each project (Battista and CSB). Second, we compare the projects' descriptions of the development of students' knowledge and reasoning about length and length measurement in detail. Third, we conclude by reflecting on the process of comparing these two approaches to describing students' developing understanding of length.

COMPARISON OF OBJECTIVES, DATA COLLECTION, AND GROUNDING FRAMEWORKS

Battista's (2001, 2012) project has constructed a set of cognitively based assessment systems that teachers can use to examine children's ways of understanding and reasoning about various core ideas in elementary school mathematics. Only the findings for length and length measurement are described in this chapter. The project pursued three interrelated objectives: (1) describing cognitive milestones and fluencies for students as they learn about length and length measurement, (2) developing a set of interview-based assessment tasks suitable for revealing students' concepts and ways of reasoning at different points in their development of understanding of length and its measurement, and (3) describing instructional suggestions that can encourage and support students to move from one level of sophistication to the next appropriate level.

CSB's project set out to describe the longitudinal development of students' understanding of length measurement across the seven-year span from pre-kindergarten through grade 5 (Clements, Sarama, & Barrett, 2007). To do this, the project pursued two related objectives: (1) providing explicit cognitive accounts of students' ways of coordinating their conceptual and strategic knowledge through four years of instruction on length measurement concepts, and (2) designing and establishing increasingly demanding sets of instructional tasks matched with the levels of knowledge and strategy.

Thus, the Battista project focused on the development and use of an extensive set of task-based interviews and mini-teaching experiments with a cross-section of elementary school students. It also focused on creating a broad account of elementary students' development of increasingly sophisticated reasoning about length, with an emphasis on the cognitive underpinnings of students' reasoning and shifts in reasoning. In contrast, the CSB project focused on longitudinal accounts of specific students in a cohort, using tutoring sessions as a context for designing instructional tasks that prompted increases in sophistication of reasoning and knowledge of length measurement. This project has emphasized the integral use of assessment and instruction to check theoretical accounts of increasing levels of knowledge.

The projects collected data differently. Battista's project (2001) focused on cross-sectional data obtained through one-on-one task-based clinical interviews and mini-teaching experiments. The data were collected in a multifaceted, iterative "assess-analyze-revise" scheme. In each of three assessment cycles, implemented in three different years, each student in an assessment cohort was interviewed five to six times over a period of several weeks, with adjustments and follow-ups to interviews made as suggested by ongoing analysis of the data. CSB's ongoing project (Clements et al., 2007) has focused on two sets of data, primarily a longitudinal account of student learning over a four-year teaching experiment, and another set of cross-sectional data collected with verbal interview protocols or written instruments at the beginning and again at the end of four years.

Both groups of researchers work within a general constructivist conceptual framework, focusing on how students construct new mathematical meanings from the relevant meanings they currently possess. However, Battista describes his conceptual framework as a set of levels of sophistication, whereas CSB explicitly describe their conceptual framework as a learning trajectory. That is, Battista's levels-of-sophistication framework focuses on describing students' construction of meaning for length in general, over a wide variety of curricula. He construes his set of levels, similar to one definition of learning progressions: "descriptions of successively more sophisticated ways of reasoning within a content domain based on research syntheses and conceptual analyses" (Smith, Wiser, Anderson, & Krajcik, 2006, p. 1). He envisions his levels as the cognitive terrain for student learning trajectories—actual trajectories of individual students, as well as hypothesized trajectories for groups of students in a specific curriculum. In contrast, CSB's learning trajectory framework focuses more closely on the coordination of a specific instructional sequence with successive levels of reasoning and understanding, seeing a learning trajectory as including "the learning goal, the learning activities, and the thinking and learning in which students might engage" (Clements & Sarama, 2004, p. 84). Both frameworks, however, assume that their theoretical descriptions of a learning sequence are tied to, and must interact with, instruction (as opposed to being general instruction-free stages of development).

Another set of common grounding assumptions arises from the procedures used to create the two descriptions of students' development of reasoning about length and length measurement. Both efforts included the following typical steps relevant to the production of their descriptions:

1. Focus on a central and fundamental concept (length) for the discipline (mathematics)
2. Describe a sequence of increasingly sophisticated ways that students understand and reason about a concept over time
3. Identify major shifts in reasoning and sophistication, complemented by connective sets of minor shifts
4. Describe typical cognitive obstacles or stumbling points along the sequence
5. Integrate psychological/developmental and disciplinary analyses of sophistication in thinking
6. Address appropriate instructional actions along the sequence2
7. Base the analyses of student thinking and actions on empirical investigation (Confrey, 2009)

Both research groups build on extant research addressing central conceptual components of students' reasoning about length (Lehrer, 2003). These include: (1) identifying the attribute of length to compare and quantify, (2) building appropriate length units with conservation, (3) partitioning/iterating objects to generate quantity, (4) composing and coordinating groups of units in structures cumulatively, and (5) iterating units in correspondence with number schemes, including an assignment for zero.

A substantive difference between the projects is that Battista addresses instruction (Step 6, above) for students at each level as a separate issue, whereas CSB embed instruction within the levels of their trajectory (cf. Sarama & Clements, 2009, pp. 17–24). Indeed, in his research, Battista has taken instruction to be students' experience of all teaching about length by teachers, curriculum materials, parents, and so on, whereas CSB have focused on particular interactions with the case students within the context of a particular cyclical set of teaching sessions across a four-year span. Thus, there is a difference in scope, variability, and generality of the instructional "treatments" leading to the two different descriptions of student learning.

DETAILED DESCRIPTIONS OF THE TWO APPROACHES FOR DESCRIBING DEVELOPMENT

We first describe a modified version of the learning trajectory developed by Clements, Sarama and Barrett (2007) in Table 4.1. We then describe the levels of sophistication developed by Battista (2003, 2006, 2009), represented in Figures 4.3a and 4.3b.

The CSB length trajectory consists of a specific hierarchical sequence of 10 levels (see Table 4.1). These researchers characterize each successive level as increasingly sophisticated and integrative of prior levels, and anticipate fallback between levels (Clements & Sarama, 2007). The higher layers indicate increasing shifts toward more abstract patterns of reasoning that become dominant, yet the child retains even the earliest layers although they are decreasingly engaged over time (Sarama & Clements, 2009). This trajectory is codeveloped in tandem with a specific, idealized sequence of instructional activities; it is based on empirical accounts of children's reasoning as they were guided through longitudinal teaching experiments with individuals, or groups and classes of children (Barrett & Clements, 2003; Barrett et al., 2012; Barrett, Jones, Thornton, & Dickson, 2003; Cullen, 2009; McCool, 2009; Sarama & Clements, 2009). A portion of this trajectory has been shown to reflect a viable hierarchical sequence by statistical analysis using a Rasch model with levels pertinent to pre-kindergarten through grade 2 performance (Szilagyi, 2007). Here we use the term *viable* to indicate that the empirical analysis of levels specified in the trajectory have proven stable, coherent, and consistent for students who were assigned to levels based on their performance over time.

Battista's levels of sophistication in students' reasoning about length are briefly summarized in Figures 4.3a and 4.3b. They were developed in his cognition-based assessment (CBA) project. A complete set of CBA levels for a topic

1. begins with a description of the informal, pre-instructional reasoning that students typically possess about a topic
2. concludes with a description of the formal mathematical version of the topic
3. indicates cognitive plateaus reached by students in moving from (a) to (b)
4. includes descriptions of cognitive obstacles students face, descriptions of students' conceptualizations, reasoning, and strategies at

TABLE 4.1. Learning Trajectory for Length Developed by Clements, Sarama, and Barrett (2007)

Level	Observable Actions	Internal Actions on Objects	Instructional Tasks
Pre-Length Quantity Recognition	Task: Describe the length of this string. [Show student a 26 inch string] Response: "This is long. Everything straight is long. If it's not straight, it can't be long."	Compares size generally without particular dimensions. Confounds length, area and volume, etc.	Order sets of objects by various attributes: model the use of terms: long, tall, high, heavy; compare same object using different attributes.
Length Quantity Recognition	Ask students to identify the height of a person: "I'm tall, see!"	Identifies length and distance as attributes. Uses categories but not comparative length.	Challenge students to compare several "tall" persons and find which one is taller than the others. Prompt comparison
Length Direct Comparison	Have students compare sticks or other objects to find the longer: Stands two sticks up next to each other on a table and says, "This one's bigger." When asked to compare objects that are not adjacent and lined up at one edge, the student gives up as they are "too far away!"	Physically aligns two objects to compare. A ruler may be used to compare directly to objects. Uses terms: long, longer, longest.	Ask children to show which of them are taller than others all at once by making an ordered line. Compare objects that are not adjacent.
Indirect Length Comparison: ILC	Children can solve everyday tasks: Is a doorway wide enough for a table to go through? Can compare the length of two objects with a piece of string as long as one of the two. When asked to measure a line segment: moves finger along a line segment, saying 10, 20, 30, 31, 32 (not consistently for units)	A mental image of a particular length can be built, maintained, and manipulated. With the immediate perceptual support of some of the objects, such images can be compared. A counting scheme operates on an intuitive unit of space or of movement.	Challenge students to compare two objects that are not directly adjacent, providing a third object that is shorter than the shorter of the two objects.
End-to-End Accumulation: EtoE	How many green strips long is the pink paper strip? The student succeeds, ignoring extras. paper strip If the green strips are scattered, the student may well succeed in aligning exactly 8 strips and reporting 8 strips. Needs all strips.	An implicit concept that lengths can be composed as repetitions of shorter lengths underlies a scheme of laying lengths end to end (at first, only for small numbers of units). The scheme improves by attending more explicitly to covering distance or composing a length with parts.	Pose tasks with fewer unit pieces available than needed to span the object; remove them from sight in a slow sequence. Cover a portion of an object and ask students to lay unit objects end to end along the partly obscured object to find its length. Can we predict length without the unit objects?

Level	Observable Actions	Internal Actions on Objects	Instructional Tasks
Unit Repeating and Relating: URR	Uses rulers with minimal guidance and some success. Measures a book's length accurately with a ruler if the ruler is aligned to the zero point. Otherwise, reports number at right most end: 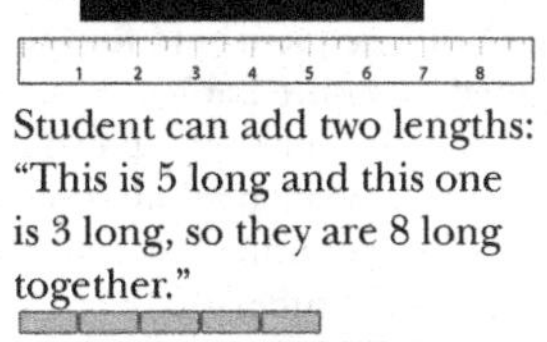 Student can add two lengths: "This is 5 long and this one is 3 long, so they are 8 long together." paper strip Students can imagine a train of units being extended: The students can start from 5, and iterate one or more green strips to find the length of 8 strips.	Action schemes include the ability to iterate a mental unit along a perceptually available object. The image of each placement can be maintained while the physical unit is moved to the next iterative position. With the support of a perceptual context, scheme can predict that fewer larger units will be required to measure an object's length. These action schemes allow counting-all addition schemes to help measure.	Use rulers and check for repeatability of measures, especially with fragments of rulers (a constraint), where neither end of the object is next to the zero point on the ruler. Present students with situations that will conflict with their sense of conservation of length (i.e., have them measure a 4-inch strip with a broken ruler and a 5-inch strip with an unbroken ruler. Student is likely to report a measure of 5 inches for both. Allow the student to compare the two strips directly).
Consistent Length Measuring : CLM (for straight paths)	Find the length of a 3.14 meter-long rope: "I used a meter stick three times, then there was a little left over. So, I lined it up from 0 and found 14 centimeters. So, it's 3 meters, 14 centimeters in all." Can also answer *broken ruler* tasks. Students can compare two paths by relating the parts in a 1:1 relation of successive parts, although they may fail to coordinate side lengths along complex paths, especially polygon perimeter: A student might report a perimeter as 25 units even though the sides added to 26 units in all; the student had marked the initial point 0 and considered it "counted already" when finding the perimeter by counting through the entire path, so she stopped at 25, rather than 26.	The length scheme has additional hierarchical components, including the ability simultaneously to image and conceive of an object's length as a total extent and a composition of units. This scheme adds constraints for equal-length units and, with rulers, on use of a zero point. Units themselves can be partitioned to increase precision.	Challenge students to integrate side lengths and perimeter measures in problematic situations, based on the conservation of both whole and parts within a figure (e.g., see the drawing where a student would "fail to coordinate side lengths" example in this row.

(continued)

TABLE 4.1. Continued

Level	Observable Actions	Internal Actions on Objects	Instructional Tasks
Conceptual Ruler Measuring: CR	Student can: *find the length of an object that is not partitioned* (or an open gap to scan as a distance). To measure across a room, a student responds: "I imagine one meter stick after another along the edge of the room. That's how I estimated the room's length is 9 meters." Students use explicit strategies to estimate lengths, including developing benchmarks (e.g., a 6-inch dollar bill). Students are not yet able to draw and coordinate a comprehensive set of cases.	Interiorization of the length scheme allows mental partitioning of a length into a given number of equal-length parts or the mental estimation of length by projecting an imaged until onto present or imagined objects.	Challenge students to integrate several sub-paths along a route to find an overall path length, coordinating the units across several iterated sequences to find the entire path and its relation to other complex paths. Challenge with erroneous counting for combinations of sub-paths.
Integrated Conceptual Path Measuring: ICPM	Computes perimeter of a polygon, including complex cases and determines sides from perimeter. Students can find several related cases of polygons with the same perimeter, or the same area, and relate those cases to one another by logical comparison. Students can make changes in one part of a figure and adjust other sides to compensate for changes of length to maintain the fixed overall path length, yet they *do not yet manage a series* of such changes in a systematic way across multiple figures to meet the overall challenge of a comprehensive set. Student can *rearrange portions of a path A to B* without number measures for the parts involved and argue that both paths are the same length, based on knowing rectangles have congruent opposite sides.	Interiorized scheme is extended to incorporate multiple units and collections of units to compare related paths or perimeters from sets of polygons. This improved scheme integrates sets of units along each section of a path object.	Students should be challenged to relate multiple cases of paths in a dynamic way, extending prior operations on a single path, to show how the dynamic interactions among paths can be used to indicate the close modifications of one portion of a path in a complementary relation to another portion of a path.

Level	Observable Actions	Internal Actions on Objects	Instructional Tasks
Coordinate and Integrate Abstract Measuring with Derived Units: AMDU	*Find and sketch all the rectangles with a specified perimeter. OR, find all the rectangles with a specified area. OR, find all the triangles with a specified perimeter.* Students can work with computed measures (derived units) of rates, such as miles per hour, or feet per second, and make conversions to manage the rates in ways that are proportionally consistent.	Well developed and interiorized scheme for linear paths allows coordination of collections of units, collections of units of units, and collections of entire paths (complex, bent paths). Coordination among length scheme and scheme for numeric ratios allows spatial operations on complex measures with derived units as quantities.	Students may be challenged to justify the choice of the significant figures in science contexts, and may be expected to argue for acceptable computational error, based on the precision of instruments used to find information and make claims about rates of processes or rate of speed.

various points in development, and descriptions of cognitive processes underlying students' various conceptualizations and reasoning

This plateau structure was illustrated in Figure 4.1 and was conceptualized as the "cognitive terrain" that students must ascend if they are to attain understanding of a major mathematical idea. In Battista's levels-of-sophistication scheme, students might take different individual trajectories through the cognitive terrain in developing reasoning about length, depending on the curriculum and the student. Some students might skip certain levels, especially levels that describe incorrect reasoning (although students may pass through such levels so quickly that it is unlikely that they can be observed doing so). Battista's levels (see Figures 4.3a and 4.3b) do not describe instruction. Other CBA work describes kinds of instruction that are likely to help students ascend to the next reasonable level (Battista, 2012).

Similarities and Differences

Both research groups ground their descriptions of the development of student reasoning in observable student actions. We illustrate this emphasis on observable student actions with an example, drawn from the CSB project, that characterized shifts in students' strategies and developing knowledge across a 22-month span from grade 2 (February) to grade 4 (November). The team engaged in a cyclical set of teaching experiment sessions with eight students, and summarized the progress of one student as a prototypical case (Student A).

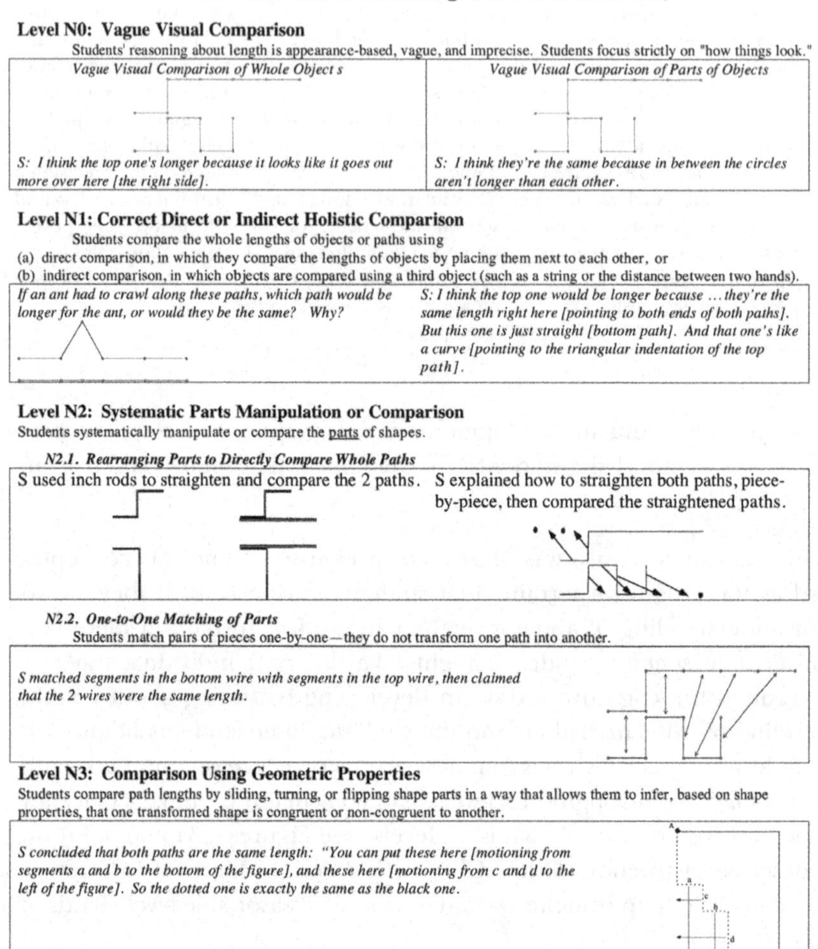

FIGURE 4.3a. Battista's nonmeasurement levels of student reasoning about length. *Note:* Battista's levels of sophistication in student reasoning about length (development of nonmeasurement, or nonnumerical, reasoning).

Measurement Reasoning [USES NUMBERS]

Measuring the length of an object consists of determining the *number* of fixed unit lengths that fit end-to-end along the object, with no gaps or overlaps.

Level M0: Use of Numbers Unconnected to Appropriate Unit Iteration

Students use numbers to describe lengths; however, their numbers do not represent iterations of a unit length.

S: [Counts 12 dots on A and 7 dots on B] A is longer.

A

B

Level M1: Incorrect Unit Iteration

Students iterate non-length units, or their iterations of length units contain gaps, overlaps, or different lengths.

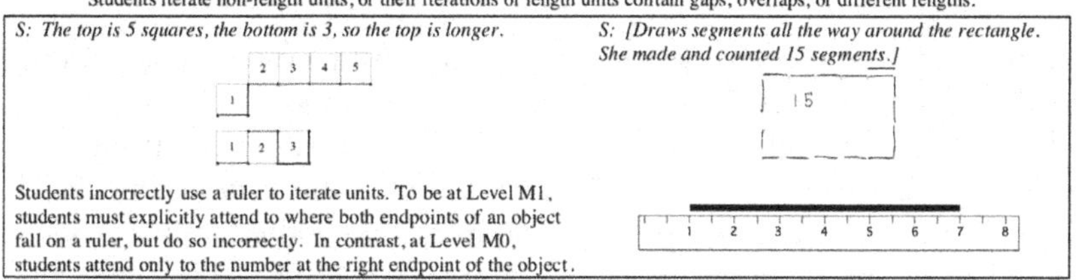

S: The top is 5 squares, the bottom is 3, so the top is longer.

S: [Draws segments all the way around the rectangle. She made and counted 15 segments.]

Students incorrectly use a ruler to iterate units. To be at Level M1, students must explicitly attend to where both endpoints of an object fall on a ruler, but do so incorrectly. In contrast, at Level M0, students attend only to the number at the right endpoint of the object.

Level M2. Correct Iteration of All Units

As students iterate unit-lengths along an object, they properly coordinate the position of each unit length with the position of the unit that precedes it so that gaps, overlaps, and variations in unit-lengths are eliminated.

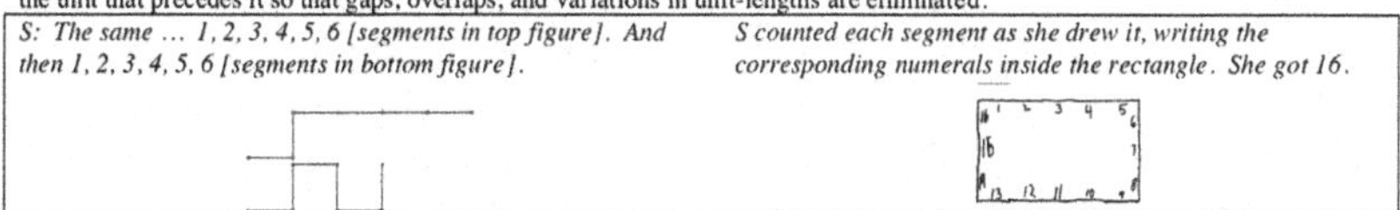

S: The same ... 1, 2, 3, 4, 5, 6 [segments in top figure]. And then 1, 2, 3, 4, 5, 6 [segments in bottom figure].

S counted each segment as she drew it, writing the corresponding numerals inside the rectangle. She got 16.

Level M3: Correct Operation on Partial Iterations

Students determine length measurements by performing logical or arithmetic operations or actions on the results of their previous iterations of unit lengths. They do not iterate all the units in a measurement.

S counted the 5 segments between the hash marks on the top of the rectangle, then said that since the bottom was the same as the top, it would also be 5. He then counted 3 on the left side, then said that the left side was equal to the right side, so the right side would be 3. S then said, "3+3=6 and 5+5=10. So it takes 16 black rods."

After counting unit lengths for two sides of the rectangle, for the other two sides, S replaced iteration with logical inference. He then operated on his results by adding them all together.

Level M4: Correct/Meaningful Operation on Measurement Numbers—No Units Visible or Iteration

Students numerically or inferentially operate on length measurements *without iterating unit lengths.*

S: The total [horizontal] length of the top is 20 + 40 = 60. So the whole [horizontal] bottom has to be 60. The part of the bottom we know is 10 + 20 = 30. So the "question mark" side is 60 – 30 = 30.

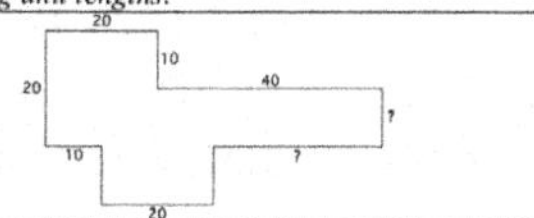

Level M5: Understanding and Using Procedures/Formulas for Perimeter

Students understand and can use perimeter formulas. They can use variables to reason about length; they don't have to use actual numbers. Students develop, use, and justify procedures or formulas to find the perimeter (distance around) of polygons. For example, students develop the formula that the perimeter of a rectangle or a parallelogram is 2 times the sum of the lengths of two adjacent sides. $P = 2(x + y)$

FIGURE 4.3b. Battista's measurement levels of student reasoning about length. *Note:* Battista's levels of sophistication in student reasoning about length (development of measurement, or numerical, reasoning).

TABLE 4.2. Strategy Levels of Length Trajectory Exhibited by Student A During a 22-Month Period

Observation Month and Grade	Feb. Gr. 2	Mar. Gr. 2	Apr. Gr. 2	May Gr. 2	Oct. Gr. 3	Feb. Gr. 3	May Gr. 3	Sep. Gr. 4	Oct. Gr. 4
Strategy Levels exhibited	URR EE IDC	URR	(CLM) URR	CLM URR	CLM	URR	(CR) CLM	CR CLM	(ICPM) CR CLM

The teaching experiment methodology employed by CSB comprised a cyclical process of monthly tutoring sessions and revision of a cognitive model of the students' developmental level. The CSB team iteratively developed a predictive model of a student's reasoning and actions, checked the fit of the model with the responses of students to assigned tasks, revised the model, made predictions about reasoning and actions given a particular set of tasks, and then prepared the next tutoring session (cf. Steffe & Thompson, 2000). For example, a student might not respond to the tasks as predicted by the model of that student's thinking. If the student was unsuccessful with these tasks, the researchers would have revised their model of the student's thinking by assigning that student to a prior, less sophisticated level in the trajectory and by finding tasks to fit that new model.

The growth of one student along the learning trajectory is reported in Table 4.2 and pictured in Figure 4.4. Student A exhibited unit repeating and relating (URR) strategies in February 2008 (see left-most observation points on graph, indicated by *Os* near the vertical axis). That is, she was not able to resolve broken ruler tasks requiring the reassignment of a zero point on a measuring scale, but she was able to repeatedly shift a single unit, scooting the unit precisely through its own length while counting successive placements along the object until she traversed the entire object length (without a ruler). By the third month of grade 4, Student A was using strategies that fit the integrated conceptual path measuring (ICPM) level. That is, she was able to coordinate and organize several related cases of rectangles with perimeter of 24 that constituted a complex solution set to a fixed perimeter task (find all rectangles with perimeter of 24).

Differences

The first obvious difference is that Battista differentiates nonmeasurement reasoning about length (reasoning without numbers) from measurement reasoning (reasoning with numbers), whereas CSB argue that measurement includes early attempts at reasoning about length that do not yet involve counting units. Battista argues that measurement occurs only when numbers are used in reasoning about unit length iteration (consistent with

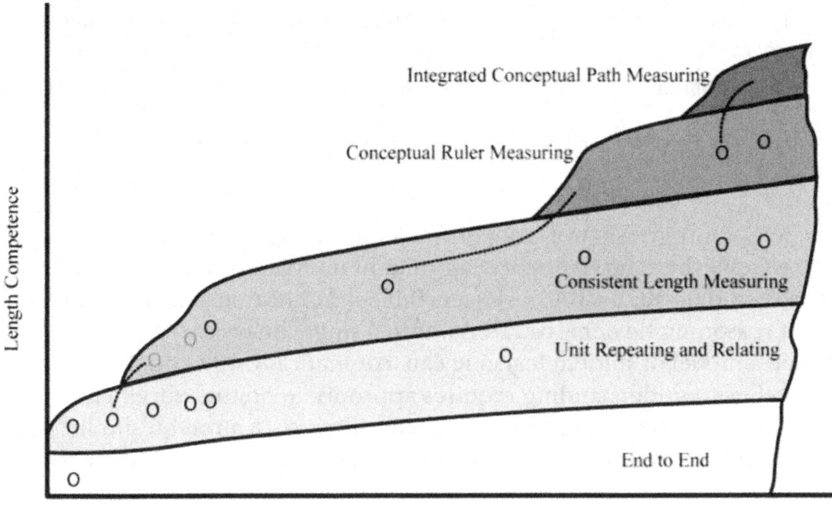

FIGURE 4.4. Student A growth along CSB length learning trajectory. *Note:* The growth of Student A along the CSB learning trajectory for length measurement over a 22-month period.

a mathematical definition of length as involving a function from the set of line segments to the set of real numbers). In contrast, CSB consider their entire trajectory as a description of the development of measurement reasoning about length. Part of this difference may involve different definitions of measurement, but the difference may also stem from different conceptions of unit length. Battista's levels suggest that students' conception of unit length occurs when such a unit is used for iteration and enumeration. CSB ascribe reasoning with a unit length to a child who has succeeded in an indirect comparison with an object smaller than either of two other objects that are themselves being compared, even if the child has no more than an intuitive notion of quantity relating to repeated images of that smallest object. From this perspective, this smallest third object would not be a numerical or iterable length unit, but instead an object that can be compared and set into a comparative relation with the other objects in a way that anticipates counting of iterated images (Piaget, Inhelder, & Szeminska, 1960).

A second major difference is that CSB locate length-reasoning strategies like direct comparison and indirect comparison prior to counting strategies such as end-to-end accumulation. Battista, on the other hand, argues that measurement and nonmeasurement reasoning about length co-occur

TABLE 4.3. Sample of Various Types of Reasoning for One Student, KG, During One Interview Session

	Problem 1	Problem 3	Problem 4	Problem 5	Problem 10	Problem SL1
Student KG	N2.1	N0	N0	N1	M2	M0, M1, M2

in the length progression of student learning, having repeatedly found co-occurrence of these two types of reasoning in interviews with students. For example, student KG's CBA work (see Table 4.3) illustrates the diversity in types of reasoning he demonstrated within a single interview.

While a model of student learning can axiomatically stipulate that learning length with understanding requires appropriate nonmeasurement reasoning to occur before measurement reasoning, such prerequisite learning does not necessarily occur in students. Many students' learning about numerical reasoning about length (and area and volume) is rote, suggesting that students can perform numerical or ruler-based procedures before they understand the underlying concepts (Battista, 2009; Smith, Sisman, Figueras, Lee, Dietiker, & Lehrer, 2008). So the difference between Battista's co-occurrence description and CSB's sequence description is more a difference between what happens in general versus what can happen in a well-designed curriculum. As we later describe, Battista agrees with CSB that in a well-designed instructional sequence, nonnumerical reasoning should precede numerical reasoning. But his levels research shows that for most students—who have not followed the careful instructional sequence described by CSB—these two types of reasoning about length frequently co-occur.

The third difference between the projects' characterizations of development is that Battista includes, *as levels*, specific ways that students reason incorrectly about length. In contrast, CSB incorporate accounts of incorrect reasoning among the features that characterize developing reasoning within appropriate levels of their learning trajectory.

CROSS-PROJECT COMPARISON OF CLAIMS ABOUT COGNITIVE DEVELOPMENT

In order to conduct a detailed comparison of Battista's and CSB's descriptions of the development of students' reasoning about length, we began by acknowledging the complex nature of the two descriptions. The two different descriptions and the implicit meaning encapsulated in the products of these research projects present a substantive categorization problem similar to the challenge of comparing apples to oranges; we decided that it was inappropriate to compare apples to oranges—that is, to compare Battista's levels of sophistication to CSB's learning trajectory. Therefore, to compare apples to apples, we examined a derived, simplified account of the CSB learning trajectory and a derived trajectory that Battista created based on his levels-development research. That is, Battista's CBA instructional theory and implementation work suggest the derivation of an instruction-based trajectory: N0, M0, N1, N2.1, N2.2, M1.4, M2, M3, M4, N3, M5 (see Figures 4.3a and 4.3b for detailed explanations of these levels). Using Battista's derived learning trajectory and CSB's simplified account of their trajectory not only makes comparison of project findings feasible, it also responds to the needs of curriculum and standards development analysts for a focused, linear account of levels of sophistication along a unidimensional trait of knowledge (cf. Baker & Kim, 2004; Martineau, 2006). Figure 4.5 presents one column from Battista's derivation of a level-based instructional sequence alongside CSB's derivation of a level-based instructional sequence.

Conceptual Analysis for Comparison

To create the comparative chart shown in Figure 4.5, we engaged in a detailed, level-by-level conceptual comparison process. Both research groups build on extant literature addressing central conceptual components of students' reasoning about length. Lehrer (2003, pp. 181–182) describes eight prominent conceptual foundations which we summarize as follows: (1) identifying the attribute of length to compare and quantify, (2) building appropriate length units with conservation, (3) partitioning/iterating objects to generate quantity, (4) composing and coordinating groups of units in structures cumulatively, (5) iterating units in correspondence with number schemes, including an assignment for zero. The first of these foundations for length reasoning fits the foundational concepts indicated by Battista Levels N0, N1, and N2 and CSB Levels 2, 3, 4, and 4-5.[3] The fit on this portion of the trajectory was nearly an exact correspondence, with minimal mismatching at Battista Level M0, Numbers Unconnected to Iteration, and the CSB End-to-End level.

We illustrate our analysis of correspondence by focusing on the conceptual connections between Battista's Level N2.2 (1-1 matching of parts) and corresponding levels in the trajectory of CSB. To find the correspondence for Battista's Level N2.2 with something in CSB's trajectory, we appealed to the conceptual underpinnings of both trajectories. The level description summary for Battista Level N2.2 reads, "students match pairs of pieces one by one—they do not transform one path into another." This indicates that the action is a correspondence from each piece of a path to each equivalent piece of another path. In the CSB trajectory, the account of student thinking for the level of consistent length measuring includes the statement, "Considers the length of a bent path as the sum of the lengths of its parts." The example includes the description that, "[students] can compare two paths by relating the parts in a 1-1 relation." Indeed, believing that "length

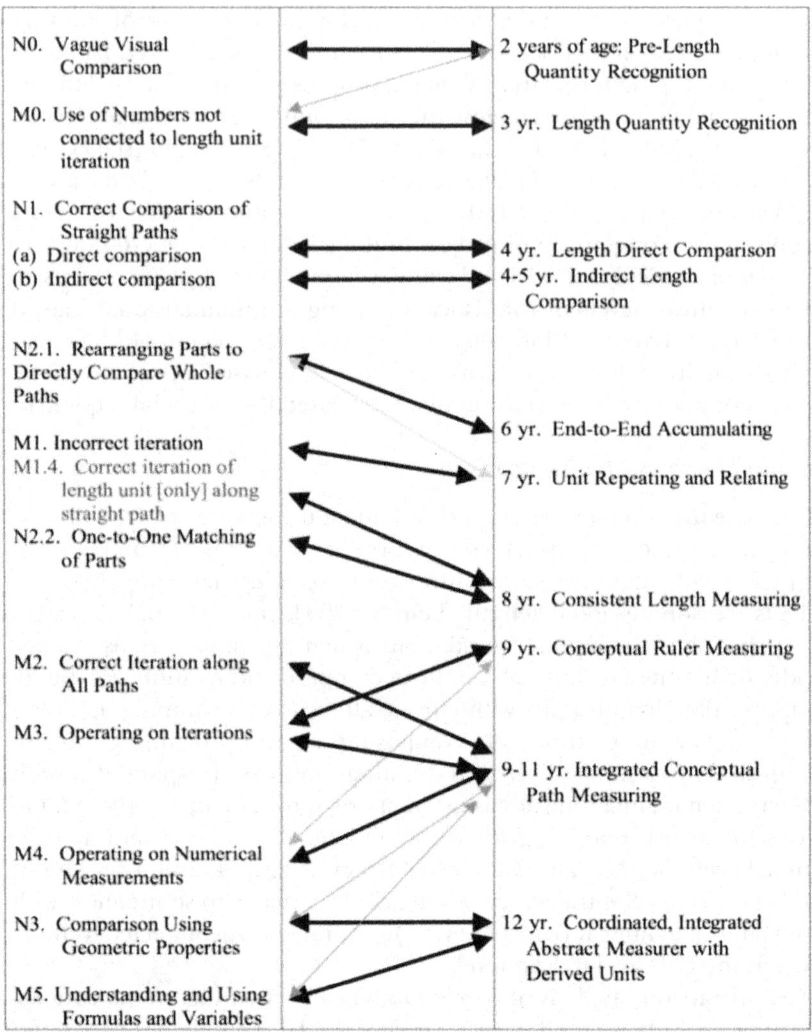

FIGURE 4.5. Mapping Battista to CSB. *Note:* A mapping between the instructional sequences of Battista (left column) and Clements, Sarama & Barrett (right column). Bold arrows indicate a clear, strong correspondence; lighter arrows indicate a weak correspondence.

of a path is the sum of the lengths of its parts" seems to implicitly rely on the belief that if a path is physically separated into parts, the length of the whole is the sum of the lengths of the parts because the parts as separated can be put into 1-1 correspondence with the parts as joined in the whole. So we included Battista's 1-1 correspondence (Level N2.2) as part of the reasoning in the CSB level consistent length measuring. This implies that it is also relevant to higher levels, specifically for conceptual ruler measuring and integrated conceptual path measuring.

Thus, as we examined the fit between the two systems, our detailed analysis and comparison of levels prompted us to renegotiate meanings by checking our attribution of mental processes at pertinent levels. This kind of detailed comparative analysis of behaviors and strategies must be based on the examination of underlying cognition if it is to be taken as a substantive comparison of trajectories. We anticipate that further trajectory research will explore this approach and others that may follow from it. The comparison outlined in Figure 4.5 emphasizes the relatively close fit between the outcomes of the projects by eliminating some of the different issues related to the treatment of instruction, and the relation between spatial, nonnumeric comparison and numerically based measurement. At the same time, this side-by-side comparison brings to light several additional important differences between the researchers' approaches to trajectories for length.

For example, sometimes the grain size differs. For instance, much happens between Battista's incorrect iteration (M1) and correct iteration (M2). In fact, CSB more elaborately describe this same developmental period through the sequence: end-to-end, unit repeating and relating, and consistent length measuring. Battista's (2003) research has also elaborated the period between his Levels M1 and M2 with the sublevels shown below. However, this elaboration does not appear in the leveled description because the simplified version of the levels shown in Figures 4.3a and 4.3b was written for teachers, not researchers:[4,5]

M1.1: Incorrect iteration with non-length unit
M1.2: Correct iteration of non-length unit along a straight path, but not non-straight paths
M1.3: Incorrect iteration with length unit
M1.4: Correct iteration of length unit along a straight path, but not non-straight paths

We note that the grain size of the sequences can vary because of the elaboration of student thinking within the research program on which the progression is based, or it may vary due to the audience for which the sequence is written.

To conceptualize our comparative analysis, we propose five types of possible correspondence between two learning trajectories: (1) Trajectories may match completely, when there is an exact correspondence of levels and of ordering of those levels, with no levels missed or skipped; (2) they may differ slightly, with minor grain-size differences that might be understood

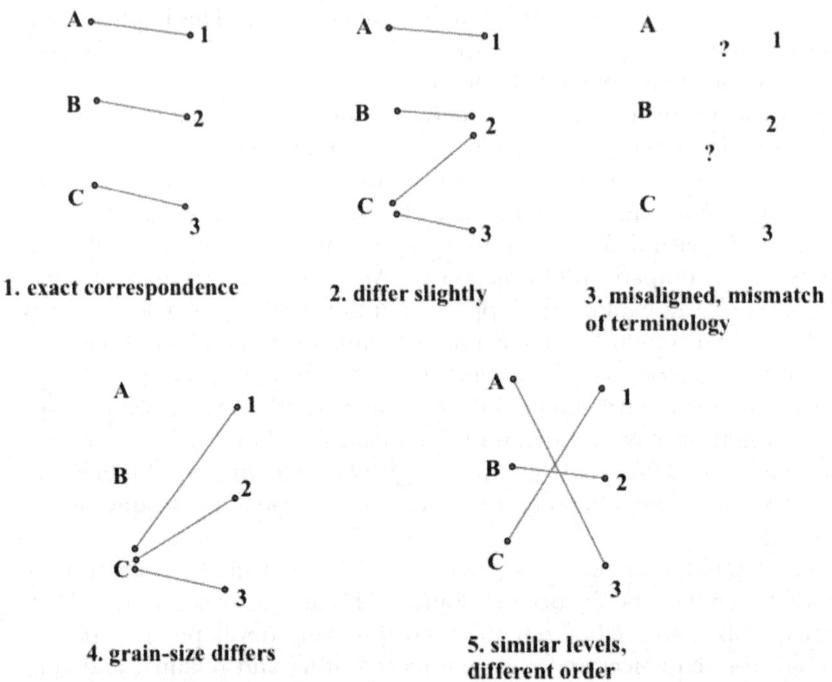

FIGURE 4.6. Types of fits among trajectories. *Note:* Several hypothetical types of fit among trajectories (e.g., Trajectories A-B-C; and 1-2-3).

by grouping two or more levels in the one trajectory so that they act as sub-levels of a single level in the other trajectory; (3) the two trajectories may indeed be misaligned, due to a fundamental difference in the underlying constructs being described by the sequences of levels (the two trajectories employ unrelated hierarchical or categorical terms, and are not merely a poor fit); (4) one may be nested within the other, as when all the levels of one trajectory are subsumed within a single level of the other, in which case the noncorresponding grain sizes of the two learning trajectories represent widely differing scope or definition of the construct(s) at hand; or (5) theoretically, they may have an inverse order relation, which we would call a complete lack of fit due to directionality, or due to different interpretations of the concepts described (see Figure 4.6). We make use of these patterns in our analysis, and we propose this as a potential grid for comparative analysis of trajectories. A sixth dimension for comparison (not shown here) is the variable use of language and constructs to describe the trajectories.

According to this comparison framework, the two trajectories described by Battista and by CSB differ slightly (Type 2) according to a comprehensive examination. While the two trajectories differ slightly overall, there are also minor grain-size differences (Type 4 above) that might be understood by grouping two or more levels in the one trajectory so that they act as sublevel details to a single level in the other trajectory (e.g., Levels M2, M3, M4, N3, and M5 match Levels 9–11). For subsets of the trajectories (e.g., M0, N1a. and N1b. matching 3, 4, and 4–5) there is an exact correspondence (Type 1). Thus we describe the overall fit as Type 2 and claim that they differ slightly.

OTHER DIFFERENCES

Elaboration of Cognitive Mechanisms

Examination of the two research groups' descriptions of trajectory levels indicates a difference in the level of detail used to describe the cognitive mechanisms underlying the levels. For instance, CSB, following a series of studies (Barrett & Clements, 2003; Barrett, Clements, Klanderman, Pennisi, & Polaki, 2006; Clements, Sarama, Barrett, Schiller, & Piyose, 2009; Szilagyi, 2007), describe their URR level as a confluence of indirect comparison reasoning, working with transitivity across visually available images and iterative actions related to each other and to the counting sequence. When these subconcepts for length coalesce, the student is able to reason in a way that fits the URR level. Battista's levels, *as written for teachers*, do not include such an account of subprocesses as necessary inputs to the level of correct iteration along straight paths (M1.4). Nevertheless, Battista's ongoing research suggests that students are able to iterate units in a way similar to CSB's URR level when they correctly choose a unit with the appropriate attribute (a unit *length*), and their unit-length iteration has the appropriate properties—iteration without gaps, overlaps, or variations in length. Understanding the properties of unit-length iteration, in turn, depends on the cognitive processes of coordination of scanning motions, accumulation of repeated unit-scanning motions into composite units, spatial structuring of the set of unit iterations, and coordination of spatial unit iteration with numeric counting (Battista, 2007). In all, the two projects incorporate similar collections of cognitive subprocesses to explain unit iteration, a Type 2 correspondence; they differ however by reporting the processes with different grain-size analysis in different contexts, a Type 4 correspondence.

Empirical Methods

The empirical methods used by the two research groups have probably contributed to some of the differences in the researchers' descriptions of the development of length reasoning. For example, CSB engaged in a longitudinal examination of data, using the iterative teaching experiment approach, which involved constructing particular curricular and instructional experiences and focused on the changes in particular students' measurement thinking and strategies for length throughout a four-year data cycle. In contrast, Battista studied student reasoning in the context of classrooms in which the curriculum and instructional experiences were already established and not modified by the researcher. He conducted several rounds of cross-sectional student interviews of students in grades 1–5. Each student was interviewed several times on numerous length tasks over several weeks, often with attempts to move students from one level to the next. Nonetheless, student interviews for both groups involved students working and reflecting on rich tasks that often promote changes in student reasoning, and thus both research groups have observed microgenetic changes in students' reasoning during single interviews.

HIGHLIGHTING THE ISSUES IN COMPARING DESCRIPTIONS: OUR REFLECTIONS

Our comparison of the two developmental sequences for length suggests several problems that are likely to arise whenever such descriptions are compared. We have observed differences in

1. the definitions of, and different researchers' preferences for, the terms "learning trajectory," "learning progression," and "levels of sophistication"
2. how students' learning and development are described and conceptualized, (consequently requiring deep conceptual analyses of the meanings in particular level descriptions)
3. the grain size of levels
4. attention to description of underlying cognitive mechanisms
5. audiences for whom the levels are written
6. the assessment tasks used by the research groups, which can elicit different kinds of student reasoning
7. the methods of data collection
8. the connection between sequences of reasoning levels and instruction

We elaborate here on two items in particular, Items 6 and 8. Item 6, assessment tasks that elicit different reasoning outcomes, can be illustrated by comparing different ways the two research groups assessed students' abstraction of unit iteration. Battista's assessment of M3, operating on iterations, is based on several tasks, one of which employs a rectangle with hash-mark divisions along two of its adjacent edges. One process revealed by this task is students' use of properties of rectangles to curtail their unit iteration (related tasks did not require knowledge of rectangle properties, but still examined students' curtailment of iteration). CSB's assessment of conceptual ruler measuring was based on several tasks that asked students to estimate length, given a single visible unit internal to the shape, and requiring visual translations and iterations of that unit. This did not necessarily produce the same cognitive demand as the task addressing Battista's Level M3. Therefore, we examined the ICPM level in CSB for other tasks that would reveal student use of shape properties to support curtailment of the visual iteration of an internally imaged unit. One task was identified that prompted students to organize and coordinate measures across several rectangular shapes having the same perimeter. However, this task may highlight whether students organize and coordinate collections of side lengths within each perimeter and whether they maintain a consistent unit of length through the perimeter measures for several related figures. This constitutes a further cognitive action (coordinating collections of side lengths with the perimeter) that did not closely correspond to M3 from Battista.

We conclude that without observing the same students on both of these types of assessment tasks it is difficult to determine exactly how the observed processes are related. (Although collecting and analyzing this additional type of data did not fall within the scopes of these research projects, such comparisons would be productive areas for further research.) Overall then, our experience trying to establish exact correspondence among terms for comparison across our projects suggests that, without further common data collection and analysis, two trajectories from different research programs can be related but not set into exact correspondence without omitting important aspects of one or the other of the trajectories. It seems reasonable to expect extensive agreement on key elements of the trajectories and to anticipate some disagreement on less critical elements.

Item 8, above, indicates the importance of the connection between instruction and reasoning levels. This can be illustrated by various claims of Battista and CSB regarding co-occurrence of number-based and spatially based (nonnumerical) length comparisons. CSB used teaching experiments as a methodology, so their work provides opportunities to examine the ongoing interplay of nonnumerical comparative reasoning and number-based measuring as students experience a sequence of prescribed instruction that prompts both types of reasoning. Battista did not use such a highly structured, instruction-focused methodology, and found much more variation in students' ways of reasoning about length (numerical measurement reasoning versus nonnumerical reasoning by visual comparison). Again, we see the results of differing conceptual frameworks. A general learning progression that is not connected to a particular instructional sequence can look different from one connected to an instructional sequence within a trajectory.

There are additional challenges that could be discussed in comparing the approaches of the two projects. For instance, in Battista's "instruction-neutral" levels-of-sophistication approach, the wide variety of tasks given to students resulted in "level profiles"—that is, level-based descriptions of how students reasoned in general about different types of tasks. The goal was to describe in detail the overall profile of students' reasoning about length, without any assumptions about prior instruction, but in a way that could guide subsequent instruction. In contrast, when a learning trajectory is tightly connected to a particular instructional intervention, as that of CSB, one uses structured sets of instructional tasks that depend on a students' presumed position in the "learning-instruction" trajectory.

DISCUSSION

In this short chapter, we have tried to highlight problems and issues arising in comparing the work of two projects describing developmental sequences for length measurement. Our purpose has been to check for the transferability and validity of these related research outcomes describing children's learning of length measurement by comparing the trajectories. We followed the claims of both projects about the development of students' length reasoning, one conceived of as an instruction-connected learning trajectory, one conceived as levels of sophistication with associated suggestions for instruction. Although we have pointed out many differences in these levels, overall, they have much in common, suggesting the validity of their core constructs. We also provide a visual mapping among the critical components of both trajectory systems. This kind of mapping proved helpful and suggested a rubric for examining fit across two projects, or two different trajectories, as they describe the same mathematical topic.

It is important to ask what researchers in this field may want to do as multiple trajectories on related mathematical topics become available for various purposes. It seems prudent to use similar analytic processes and the framework developed here to coordinate or integrate related trajectories; such efforts might produce a generalized trajectory with greater validity or generality than its precursors. Our experience with this analysis suggests that external and internal analysis should be used in a complementary way to look for validity from both vantage points; the external vantage point helps check for the transferability of the trajectory beyond the specific research context in which it was constructed and the internal vantage point helps support the internal conceptual integrity of each trajectory within its own research project. At the same time, we also recommend further examination of trajectories from the standpoint of their intended purpose. In our case, we find that Battista's project was aimed most directly to support teachers in their practice and to guide professional development, whereas the CSB project was focused more broadly to support curricular analysis, the development of standards for curriculum, further research with students, and the design of professional development. An integration of related trajectories may be expected to support a broader range of applications, but empirical work seems needed to check the applications of such trajectories.

We have modeled a process for comparative analysis of descriptions of the development of students' learning of the concept of length, addressing similarities and differences in the approaches taken by two different research teams. This comparative analysis has included surface-level comparison, deep cognitive process comparison, methodology comparison, and, ultimately, the use of the learning-trajectory construct as common ground for the one-to-one comparison required to make general recommendations for curriculum and instruction. In fact, by the articulating instructional sequences that follow from both learning descriptions, we found that length measurement trajectories could be profitably compared and synthesized to make such recommendations. Although similar terminology was not forthcoming, the attempt to reach across the projects to understand variant uses of terms and constructs helped us clarify distinctions and similarities in our assumptions and in our findings. Finally, by attending to the different role of instructional tasks in our projects, we have highlighted the flexibility of learning trajectory research, showing the potential for finding common implications in spite of different methodological and theoretical foundations. We believe that similarities in the production of related learning trajectory systems across different research programs indicate the validity of descriptions within both trajectory systems. However, differences do not necessarily mean conclusions are invalid, but that different contexts produce different results. By examining the potential correspondence and by establishing partial fit, analysts from within the original research teams (as we show here) should expect to find reasonable fit, perhaps Type 1, 2 or 4 in our rubric (Figure 4.6).

We note too that different research programs investigating a given topic in the curriculum from other theoretical perspectives may produce further variation among the trajectory elements and this might strengthen the eventual synthesis across trajectories. For example, although Battista's levels were derived from students in a variety of curricula, and CSB in their particular instructional sequence, neither project investigated students learning in instructional sequences that are extremely different, such as Davydov (1991) and splitting approaches (Confrey & Smith, 1995; Steffe, 2002). Different descriptions might be derived from these approaches. Comparing Battista's and CSB's results with those obtained using these approaches might reveal commonalities that are more universal, or conversely, they might suggest substantive and incompatible differences. It is unlikely that there is a "best" learning trajectory. Rather, there may be several research-based learning trajectories that are highly effective in producing high-quality learning in students.

In conclusion, we suggest that tightly focused research in each field of mathematics education is critical for producing findings that can be subjected to the type of deep comparative analysis we have employed in this chapter. Ongoing research with trajectories promises to strengthen mathematics education research by providing a robust tool for coordinating outcomes across a variety of research teams; this should widen and strengthen the empirical and theoretical foundations for constructing content standards and developing curriculum (cf. Daro, Mosher & Corcoran, 2011, pp. 55–57). Lastly, we recommend further research addressing several challenges for using trajectories: How do we specify precisely the role of instructional tasks in learning trajectories? Is it possible to establish common terminology that permits straightforward comparison of trajectories for the same topic? To what extent should learning trajectory research be organized and coordinated to ensure common terminology and usage of instructional tasks so that it can be effectively synthesized for teacher development, standards development, and evaluation?

AUTHOR'S NOTE

We acknowledge and appreciate funding support from the National Science Foundation (Grants DRL 0099047, DRL 0838137, and DRL 0732217). The views and opinions expressed in this chapter represent the thinking of the authors and not necessarily that of the foundation.

NOTES

1. Battista (2012) argues that it is essential to clearly distinguish levels of sophistication and learning progressions from learning trajectories (actual and hypothetical).
2. Note, however, that although there is agreement that learning trajectories are dependent on instruction, the nature of the relationship between a specific trajectory and relevant instruction is debated (Corcoran, Mosher, & Rogat, 2009).
3. The level numbers for CSB come from approximate age in years at which the level is most likely to be observable. Thus, Level 2 is the most likely to be observed among two-year-old children. In contrast, Battista found huge variation for the ages at which different levels occurred.
4. The reasoning described in these sublevels of M1 is discussed in the CBA levels document, but not formalized into levels.
5. Battista rejected another possible sublevel between M1 and M2 (viz. correctly measuring straight paths) because his research shows that students often can do this without understanding the nature of the length unit. For example, they might iterate squares or cubes without knowing what attribute they are measuring.

REFERENCES

Baker, F. B., & Kim, S. (2004). *Item response theory: Parameter estimation techniques* (2nd ed.). New York, NY: Marcel Dekker, Inc.

Barrett, J. E., & Clements, D. H. (2003). Quantifying path length: Fourth-grade children's developing abstractions for linear measurement. *Cognition and Instruction, 21*(4).

Barrett, J. E., Clements, D. H., Klanderman, D., Pennisi, S. J., & Polaki, M. V. (2006). Students' coordination of geometric reasoning and measuring strategies on a fixed perimeter task: Developing mathematical understanding of linear measurement. *Journal for Research in Mathematics Education, 37*(3), 187–221.

Barrett, J., Clements, D., Sarama, J., Cullen, C., McCool, J., Witkowski Rumsey C., & Klanderman, D. (2012). Evaluating and improving a learning trajectory for linear measurement in elementary grades 2 and 3: A longitudinal study. *Mathematical Thinking and Learning 14*(1), 28–54.

Barrett, J. E., Jones, G. A., Thornton, C. A., & Dickson, S. (2003). Understanding children's developing strategies and concepts for length. In D. H. Clements (Ed.), *Learning and teaching measurement* (pp. 17–30). Reston, VA: National Council of Teachers of Mathematics.

Battista, M. T. (2001). *The development of a cognition-based assessment system for core mathematics concepts in grades K–8.* Kent, OH: Kent State University National Science Foundation.

Battista, M. T. (2003, July). *Levels of sophistication in elementary students reasoning about length.* Paper presented at the 27th International Group for the Psychology of Mathematics Education Conference Honolulu, Hawaii.

Battista, M. T. (2006). Understanding the development of students' thinking about length. *Teaching children mathematics, 13*(3), 140–147.

Battista, M. T. (2007). The development of geometric and spatial thinking. In F. K. Lester (Ed.), *Second handbook of research on mathematics teaching and learning* (pp. 843–908). Charlotte, NC: Information Age Publishing, Inc.

Battista, M. T. (2009, September). *Analysis of cognitively based assessment accounts of students learning of measurement.* Paper presented at the Mini-Center Working Group on Measurement, East Lansing, MI.

Battista, M. (2012). *Cognition-based assessment and teaching of geometric measurement: Building on students' reasoning.* Portsmouth, NH: Heinemann.

Clements, D. H., & Sarama, J. (2004). Hypothetical learning trajectories. *Mathematical Thinking and Learning, 6*(2), 81–90.

Clements, D. H., & Sarama, J. (2007). Early childhood mathematics learning. In F. K. Lester (Ed.), *Second handbook of research on mathematics teaching and learning* (pp. 461–555). Greenwich, CT: Information Age Publishing.

Clements, D. H., Sarama, J., & Barrett, J. E. (2007). *A longitudinal examination of children's developing knowledge of measurement.* Normal, IL and Buffalo, NY: National Science Foundation.

Clements, D. H., Sarama, J., Barrett, J. E., Schiller, J., & Piyose, N. (2009, April). *Hypothetical learning trajectory for length: A multidisciplinary study.* Paper presented at the Annual Meeting of the American Educational Research Association, San Diego, CA.

Confrey, J. (2009, August). Confrey presentation notes from Friday Institute Conference on Learning Trajectories, Raleigh, NC. Personal collection of J. E. Barrett.

Confrey, J., & Smith, E. (1995). Splitting, covariation, and their role in the development of exponential functions. *Journal for Research in Mathematics Education, 26*(1), 66–86.

Corcoran, T., Mosher, F. A., & Rogat, A. (2009). *Learning progressions in science: An evidence-based approach to reform.* New York, NY: Teachers College-Columbia University.

Cullen, C. (2009). *A comparative analysis: two representational models for length.* Unpublished doctoral dissertation, Illinois State University, Normal, IL.

Daro, P., Mosher, F. A., & Corcoran, T. (2011). *Learning trajectories in mathematics: A foundation for standards, curriculum, assessment, and instruction.* New York, NY: Teachers College-Columbia University.

Davydov, V. V. (1991). Psychological abilities of primary school children in learning mathematics. *Soviet Studies in Mathematics Education, 6,* 14–56.

Lehrer, R. (2003). Developing understanding of measurement. In J. Kilpatrick, G. Martin, & D. Schifter (Eds.), A *research companion to principles and standards for school mathematics* (pp. 179–192). Reston, VA: National Council of Teachers of Mathematics.

Martineau, J. A. (2006). Distorting value added: The use of longitudinal, vertically scaled student achievement data for growth-based, value-added accountability. *Journal of Educational and Behavioral Statistics, 31*(1), 35–62.

McCool, J. (2009). *Measurement learning trajectories: A tool for professional development.* Unpublished doctoral dissertation, Illinois State University, Normal, Illinois.

Piaget, J., Inhelder, B., & Szeminska, A. (1960). *The child's conception of geometry.* London: Routledge and Kegan Paul.

Sarama, J., & Clements, D. H. (2009). *Early childhood mathematics education research: Learning trajectories for young children.* New York, NY: Routledge.

Smith, J., Sisman, G., Figueras, H., Lee, K., Dietiker, L., & Lehrer, R. (2008, April). *Assessing curricular contributions to poor measurement learning.* Presentation at Research Presession Symposium. National Council of Teachers of Mathematics.

Smith, C. L., Wiser, M., Anderson, C. W., & Krajcik, J. (2006). Implications of research on children's learning for standards and assessment: A proposed learning progression for matter and atomic molecular theory. *Measurement: Interdisciplinary Research and Perspectives, 4*(1 & 2), 1–98.

Steffe, L. P. (2002). A new hypothesis concerning children's fractional knowledge. *Journal of Mathematical Behavior, 102,* 1–41.

Steffe, L. P., & Thompson, P. W. (2000). Teaching experiment methodology: Underlying principles and essential elements. In R. Lesh & E. Kelley (Eds.), *Handbook of research design in mathematics and science education* (pp. 267–306). Mahwah, NJ: Lawrence Erlbaum Associates.

Szilagyi, J. (2007). *Young children's understanding of length measurement: A developmental progression.* Unpublished doctoral dissertation, University at Buffalo, Buffalo, NY.

CHAPTER 5

LINKING STANDARDS AND LEARNING TRAJECTORIES

Boundary Objects and Representations

Jere Confrey and Alan Maloney

ABSTRACT

Researchers in mathematics education have conjectured that learning trajectories can support improved instructional guidance to teachers and learners. Learning trajectories can also be a boundary object that supports cooperation among diverse groups in mathematics education, by focusing on growth, over time, of student conceptual proficiency from relative novice to more complex understanding. One of the stated goals of Common Core State Standards for Mathematics (CCSS-M) is to incorporate learning progressions, or trajectories. If learning trajectories could be linked to these standards effectively, teachers could use the standards and the learning trajectories together to inform what they should teach, in relation to other topics, and in relation to the preparation their students should have for the following year. Two representations of the CCSS-M have been developed to provide more representational coherence among the standards and learning trajectories.

Learning Over Time: Learning Trajectories in Mathematics Education, pages 125–160.

These support teachers in planning and interpreting their instruction, and, more broadly, support mediation among different communities concerned with mathematics education. The representations include extensions of the standards (which contain abridged learning trajectories, at best) into the sequences of proficiency levels that comprise unabridged learning trajectories. These representations can also help define professional development priorities and frame mathematics educational research agendas into learning trajectories. They also provide a starting point for framing research and gathering data on teachers' and students' practical experience with learning trajectories and the standards (including the use of more multifaceted technological resources), to support revision of the CCSS-M over the longer term.

A HISTORICAL CONTEXT FOR STANDARDS IN MATHEMATICS EDUCATION

The concept of a *standard* has evolved over the centuries. As early as the 12th century, a s*tandard* was an authoritarian symbol, the king's standard, an ensign or banner representing the king, waving above the royal residence or battlefield, representing the king's authority or the point to which all the troops should rally. The standard was sometimes itself fought for as though it were the king himself (standard, 2007). Thus the earliest standards were essentially authoritarian—proclamations of religious and political leaders, or fighting forces, accompanied by no explanation and no justification other than the authority as warranted by the full and unchallenged status vested in the proclaimer. As the scientific revolution developed, and travel and commerce expanded dramatically, a second use of standard evolved, also connoting authority, but in this case transportable authoritative systems of measurement. In the process of *standardization* of measurement, the need for accurate measurement of all kinds—mass, weight, distance, volume, time, and so forth—and the need for instruments that facilitated coordination among different results and findings, were linked to standards.

More recently, and accompanying increased globalization and exchange of new technologies, currencies, languages, and other cultural artifacts, various entities periodically commit to *shared standards.* These are negotiated agreements on measurable constructs or proficiencies, by which progress can ostensibly be measured or accounted for (Confrey, 2007). It is this version of the term *standard* that is the focus of this chapter.

Standards have become *a means to modulate or to direct change while constraining variation* in complex systems, deriving from negotiations among authorized individuals with relevant expertise (often in relation to measurement). As with the other uses of standards described briefly above, standardization is closely linked to stabilization. Determining when and to what degree we need to act in accordance with each other, over what unit of coherence (school, district, state, or nation), over what period of time (frequency of change), and to what degree we will tolerate or support variation becomes increasingly essential in globalized societies. Authorized standard-setting bodies must technically and empirically determine what produces the best outcome—for classroom education, for example, what delivers the clearest, most reliable signals to teachers and leaders, guides practice most effectively, or can be learned and mastered by practitioners, in a reliable, fair, and valid way.

Invariably, setting standards to regulate and guide complex systems requires a bootstrapping process of setting initial standards and then making appropriate adjustments in response to feedback. The first mathematics education standards in the United States, developed in response to *A Nation at Risk* (Gardiner, 1983) by the National Council of Teachers of Mathematics, were the Curriculum and Evaluation Standards (National Council of Teachers of Mathematics, 1989). The novelty of these standards lay in the way they described mathematics to include issues of communication and problem solving, emphasized the importance of processes and conceptual understanding, and deemphasized rote memorization. The standards emerged within a policy context of systemic reform (O'Day & Smith, 1993) whose aims included the alignment of key educational elements—curriculum, instruction, teacher preparation, professional development, and assessment—to standards, while permitting local flexibility in the means to reach those goals. When Congress reauthorized the Elementary and Secondary Education Act in 1994 (Improving America's Schools Act, 1994), it made standards-based reform, including the establishment of challenging standards aligned to assessments and accountability measures, a major condition for Title I funding for the states from the U.S. Department of Education (Massell, 2008).

During the standards movement in the United States, controversies known as the "math wars" erupted, based in part on differences in professional opinions concerning:

1. The timing and the degree of the use of calculators and other technologies in schools and on assessments
2. The extent of focus on student-generated strategies vs. focus on standard algorithms
3. The use of situation-based problems to introduce and motivate the study of mathematical topics
4. The centrality of probability and statistics to school mathematics

Other debates within the math wars arose from differences in opinion about whether content should be strictly separated from pedagogy in standards. On one side of this particular debate, content standards were simply lists of the topics to be taught at each grade (Stevenson & Stigler, 1994), while on the other, it was argued that the standards should also guide teachers in selecting approaches that build the basis for student understanding (National Council of Teachers of Mathematics, 1989).

The feedback generated during the controversies prompted development of a second set of standards, Principles and Standards of School Mathematics (PSSM) (National Council of Teachers of Mathematics, 2000). These attempted to bridge the divide between the sparring communities; they succeeded modestly in depolarizing the disputes over standards. The math wars then migrated to issues of effectiveness of reform curricula rather than debates over the standards themselves (National Research Council, 2004).

The development and publication of the PSSM was followed closely by the reauthorization of the Elementary and Secondary Education Act, which became know as the No Child Left Behind (NCLB) Act of 2001, signed into law in 2002 (EdWeek Research Center, 2011). This legislation set the goal of achieving higher standards for student learning and ascertaining the extent of this achievement on a school-by-school basis through required annual tests in mathematics and English language arts, in grades 3–8. The NCLB mandated the disaggregation of student subgroup assessment data, resulting in improved transparency of subgroup progress (equity targets). The increased accountability requirements of the law included corrective as well as punitive consequences for schools when students and schools failed to meet annual yearly progress targets. Parents and others took increased interest in the standards, their relationship to curriculum, and the results of the high-stakes assessments that measured achievement (Massell, 2008).

A study of current standards found that standards are more likely to influence teachers if they (1) are consistent (aligned) among themselves so that teachers receive a coherent message; (2) are specific in their messages to teachers about what they are to teach; (3) have the authority of being both developed and promoted by experts as well as officially adopted as state educational policy; (4) are consistent with standard practice and are promoted by charismatic individuals (that is, individuals who provide leadership and motivate those who must implement the standards); (5) have power, in that compliance with the standards is rewarded while failure to comply is sanctioned; and (6) are reasonably stable, in that they are kept in place over time, before undergoing further revision (Porter, Polikoff, & Smithson, 2009).

For two and a half decades, the states developed one or more versions of their own individual sets of standards for mathematics curricula. Diverse in organization and content, the state standards were often inconsistent from state to state in content and topic coverage by year. American math standards were criticized as being "mile high and inch deep," lacking coherence and vertical alignment in mathematics goals (Schmidt, cited in National Research Council, 2008, p. 24). In 2009–2010, in recognition of the need to improve the consistency and coherence among state standards and to respond to the perceived need to streamline and improve the stringency of educational standards relative to those of economic competitors of the U.S., two organizations of state officials, the Council of Chief State School Officers (CCSSO) and the National Governors Association (NGA), undertook the Common Core State Standards Initiative, or CCSSI. The Common Core State Standards for Mathematics (CCSS-M), developed through an intensive, rapid process of writing and review, were finalized and released in June 2010. They represent negotiated agreement among a group of CCSSI-selected writers, a validation committee comprising diverse experts and various stakeholders, regarding a coherent set of mathematical standards to be offered to the various states to adopt as their own.[1] States were encouraged to adopt the CCSS by the CCSSO, NGA, U.S. Department of Education, and others, through a variety of mechanisms. To date, 45 states and the District of Columbia have adopted the CCSS (CCSSO, 2012).

Among the expressed intents of the CCSS-M was that learning progressions would be incorporated in the standards:

> [T]he "sequence of topics and performances" that is outlined in a body of mathematics standards must also respect what is known about how students learn. As Confrey (2007) pointed out, developing "sequenced obstacles and challenges for students...absent the insights about meaning that derive from careful study of learning, would be unfortunate and unwise." In recognition of this, the development of these Standards began with research-based learning progressions detailing what is known today about how students' mathematical knowledge, skill, and understanding develop over time. (CCSSO, 2010, p. 4)

Both the lead writers and the validation committee acknowledged that learning trajectory research is insufficient to support all the decisions on standards incorporated within the CCSS-M. It was, however, envisioned the CCSS would evolve and be modified over the years following the CCSS initial release, to reflect ongoing research into mathematics learning progressions: "One promise of common state standards is that over time, they will allow research on learning progressions to inform and improve the design of Standards to a much greater extent than is possible today" (National Governors Association Center for Best Practices & Council of Chief State School Officers, 2010, p. 5).

In the rest of this chapter, we discuss in more detail the linkages between learning trajectories and the CCSS-M. We comment on the role of learning trajectories as boundary objects. We introduce two representations that link the CCSS-M more explicitly to learning trajectories, improve the utility of the CCSS-M for teachers, and can serve to point the way to enhancing research agendas that develop more complete empirically based progressions of student learning that are consistent with the CCSS-M.

INTRODUCTION TO LEARNING TRAJECTORIES

Numerous educational researchers have proposed different definitions of learning trajectories. Simon (1995) introduced the term *hypothetical learning trajectory*, referring to a model of the development of students' understanding of a topic, as conjectured by individual teachers for their own classroom use, and the accompanying activities the teacher employed to further students' learning day to day. Other researchers, including Cobb, Confrey, diSessa, Lehrer, and Schauble (2003); Confrey (2006); Lehrer and Schauble (2006); Brown (1992); Clements and Sarama (2004, 2009); and others, have provided other definitions of learning trajectories/progressions, depending on the particular context in which they have worked.

For the purposes of this chapter, and to frame the discussion of the relationship of learning trajectories and mathematics learning standards, we define a learning trajectory as we did in Chapter 3 (this volume):

> A researcher-conjectured, empirically supported description of the ordered network of constructs a student encounters through instruction (i.e., activities, tasks, tools, forms of interaction and methods of evaluation), in order to move from informal ideas, through successive refinements of representation, articulation, and reflection, towards increasingly complex concepts over time. (Confrey, Maloney, Nguyen, Mojica, & Myers, 2009)

With this definition, a learning trajectories approach to standards and to classroom learning in general incorporates the following qualities, which have potential to lend coherence to the development of standards across the grades:

- Student learning is viewed as the development of expertise through the refinement, over time, of approaches and understanding to handle increasingly complex concepts and challenges.
- The complexity of learning is respected, and disentangled gradually.
- Student learning is facilitated on the basis of the cognitive resources children bring to school from informal settings and from previous classroom learning experiences ("prior knowledge").
- The "logical structure of mathematics" and "cognitive development in mathematics" are distinct, and a role is seen for each (see Chapters 1, 2, and 3, this volume).
- Instruction is recognized as central to the learning experience of children in schools—instruction, cognition, reasoning and skills are all interwoven in a dynamic environment.

LEARNING TRAJECTORIES AS BOUNDARY OBJECTS

Learning trajectories can play a role in mending some of the breaches among the communities most closely involved in mathematics education. During the previous 20 years, various communities of mathematicians, mathematics educators, teachers, and several interest groups experienced considerable distrust and communication impasses regarding the content and standards in K–12 mathematics education. In part because of the qualities of a learning trajectories approach to standards described just previously, we propose that learning trajectories can act as boundary objects, as a means of fostering communication and cooperation among different communities who have vested interests in mathematics education. Star and Griesemer's (1989) description of boundary objects is consistent with the use of learning trajectories to clarify distinctions and promote common efforts in improving mathematics education in schools. In Chapter 2, Lehrer et al. (this volume) have already mentioned generating boundary objects in their research on one particular learning progression, finding them important to mediate "coordination and conflict among communities" comprising teachers, learning researchers, and assessment researchers (see Chapter 2, this volume, Abstract). Here we invoke learning trajectories and representations of learning trajectories and standards collectively as boundary objects that can be shared among an even broader set of communities.

The communities with interests in mathematics education and its learning standards are analogous to the various scientific social worlds described by Star and Griesemer (1989). Different groups or stakeholders share an overarching goal of improving the mathematical education of children. To make sustained progress toward this goal, the various groups require some common objects or norms, or some standardized forms of communication. Under the umbrella of this overarching goal, multiple groups play different and sometimes overlapping roles: teachers, students, professional development specialists, mathematics education researchers, mathematicians, school and district administrators, policymakers, assessment developers, and parents. Translating among these groups and promoting cooperation can be, we believe, facilitated with well-chosen boundary objects. Learning trajectories are, we argue, the kind of boundary object that can foster cooperation and progress among many of these groups by helping to maintain a focus on the development of children's mathematical reasoning and skills.

Star and Griesemer's (1989) development of the construct *boundary object* was instrumental in explaining how various "social worlds"—all of which have a stake or interest in, but heterogeneous viewpoints on, a particular issue or institution—can cooperate productively despite the absence of perfectly aligned consensus on the issue or institution. Boundary objects permit meanings to be taken up in different ways by different communities, while still retaining commonality among the different communities about salient features:

> Boundary objects are objects, which are both plastic enough to adapt to local needs and constraints of the several parties employing them, yet robust enough to maintain a common identity across sites. They are weakly structured in common use, and become strongly structured in individual-site use. They may be abstract or concrete. They have different meanings in different social worlds but their structure is common enough to more than one world to make them recognizable means of translation. The creation and management of boundary objects is key in developing and maintaining coherence across intersecting social worlds. (Star & Griesemer, 1989, p. 393)

The boundary object concept was set in the context of "translation" among social worlds, which is essential if a degree of coherence and cooperation is to emerge from groups with multiple, even divergent, points of view who must somehow work productively together. An important corollary of this kind of cooperation among heterogeneous actors is the reduction of local uncertainty in communication and, in essence, the development or emergence of standardization of methods of inquiry, or of forms of evidence and communication, both within and between social worlds or communities.

The boundary object should be thought of not so much as a delineation *between* spaces—not, for instance, a fence to keep different communities separated—but rather a *shared* space *among* social worlds or communities. The term *object* is used in the sense of something that people act toward and with, but certainly does not necessarily connote a physical object (Star, 2010). The boundary object is something that people use in different ways in the "shared" space between and among communities that have different particular interests, to facilitate communication, translation, and cooperative action among them.

This description implies, however, that different communities may define or use the boundary object differently. Star (2010) refers to the "interpretive flexibility" provided by boundary object, in the sense that imprecision in defining or describing a boundary object may actually *facilitate* communication and cooperation among the involved communities: overspecifying or rigidly defining the object can undermine cooperation, so a degree of imprecision can serve as a kind of flexibility that promotes cooperation. At the same time, within a particular community, the boundary object may need to be defined with greater detail and more systematically to permit further progress within that community (through routinization of work, for instance) while still being consistent with cooperation among the communities.

It would be naïve to expect complete reconciliation of longstanding disagreements between such groups—witness the report of the National Mathematics Advisory Panel (2008) and the ensuing vigorous critiques in a special volume of *Educational Researcher* (Kelly, 2008). However, well-chosen boundary objects can lead to increased agreement and cooperation among such groups across broad areas of mathematical learning and content, instead of failure to understand each other and increasing friction or animosity. The structure of learning trajectories incorporates goals held by both mathematicians and mathematics educators, not to mention teachers and other groups: the growth in student understanding of and reasoning with key concepts of mathematics and a research base that informs instructional design and decision making to achieve that growth. Both mathematics educators and mathematicians can use learning trajectories as an empirically based referent for supporting children's mathematical cognitive development over time. Appropriately designed representations of the relationship between learning progressions and standards can provide a common, easily interpreted display that facilitates translation of the standards and the learning progressions among multiple communities.

Under NCLB, the accountability system was based on bookends of standards and high-stakes assessment, but it largely neglected the instructional core and any coherence between standards and those high-stakes assessments. Learning trajectories can provide a common framework for student conceptual growth and thus support coherence among standards, instruction, and assessment (Figure 5.1). Professional development, classroom instructional practices, curriculum development, and assessment practices, if focused on growth of student learning, could be guided and made more mutually coherent by reference to learning trajectories. Based on empirical study of student learning, learning trajectories can help identify when student conceptual growth entails *intermediate steps*, and when rigorous and precise definitions are needed and/or appropriate. If a learning trajectory is less well developed empirically, mathematics educators, teachers, and mathematicians can identify areas for discussion and research.

In different contexts, the particular use of the learning trajectory construct may differ. Teachers' increasing familiarity with learning trajectories for particular constructs may lead to increased implicit understanding of and support for student contributions to discourse. Teachers could also refer explicitly to the learning trajectories to interpret student work, design formative assessment, or plan instruction (see Chapter 8, this volume). The levels of a trajectory, with carefully defined cognitive behaviors and assessment outcome spaces, are critical underpinnings for various assessments (see Chapters 1, 2, and 3, this volume; Confrey & Maloney, 2012). Curriculum based on learning trajectories may be structured so that daily lessons are built from and indexed to levels of one or more learning trajectories, with supporting materials that guide the teacher's expectations of likely student responses to particular instructional experiences (see Chapter 1, this volume). Alternatively, the ordering of constructs in learning trajectories could influence the design of projects and classroom activities, even while retaining flexibility in the order in which students encounter constructs from a particular learning trajectory. Possible research foci that remain largely untapped include the ways that multiple learning trajectories share particular proficiencies (nodes) and the ways that learning trajectories that embody different but related concepts interact and reinforce one another to the benefit of students' cognitive growth.

Learning trajectories—in the role of boundary objects—can support an unprecedented degree of coherence among standards, assessment, and accountability and alignment with instruction and curriculum. In asserting that the CCSS-M standards should be based on learning progressions insofar as research supports it and stating its intent that subsequent revisions of the CCSS-M incorporate further learning progressions research, the CCSSI took a major step toward supporting coherence of the overall education system (standards, curriculum, classroom discourse, and assessment). In the remainder of this chapter, we describe two representations we have designed to support use of learning trajectories to interpret standards and support "translation" among different stakeholders in mathematics education.

REPRESENTING STANDARDS TO REFLECT STUDENT LEARNING DEVELOPMENT

While standards form a kind of touchstone for the development of curriculum and assessment, are they something that practitioners can readily use in their practice, to guide their instruction? We have posed the question, "Have you ever closely read the standards [either the CCSS-M or current state math standards]?" to audiences at numerous workshops. The question is typically met with nods, but also many knowing chuckles and sarcastic comments ("Oh, sure..."). Teachers, administrators, and policy experts readily acknowledge the importance of standards, but most also readily agree that they are difficult to read, awkward, and even impenetrable, as a tool to directly inform instruction and planning.

The Common Core State Standards represent a major investment of expertise, time, and energy in identifying and articulating their topics and coverage. The writers of the CCSS-M intended to incorporate research on learning progressions in the standards. The writers state, "Ideally...each standard in this document might have been phrased in the form, 'Students who already know...should next come to learn....' But at present this approach is unrealistic—not least because existing education research cannot specify all such learning pathways" (National Governors Association Center

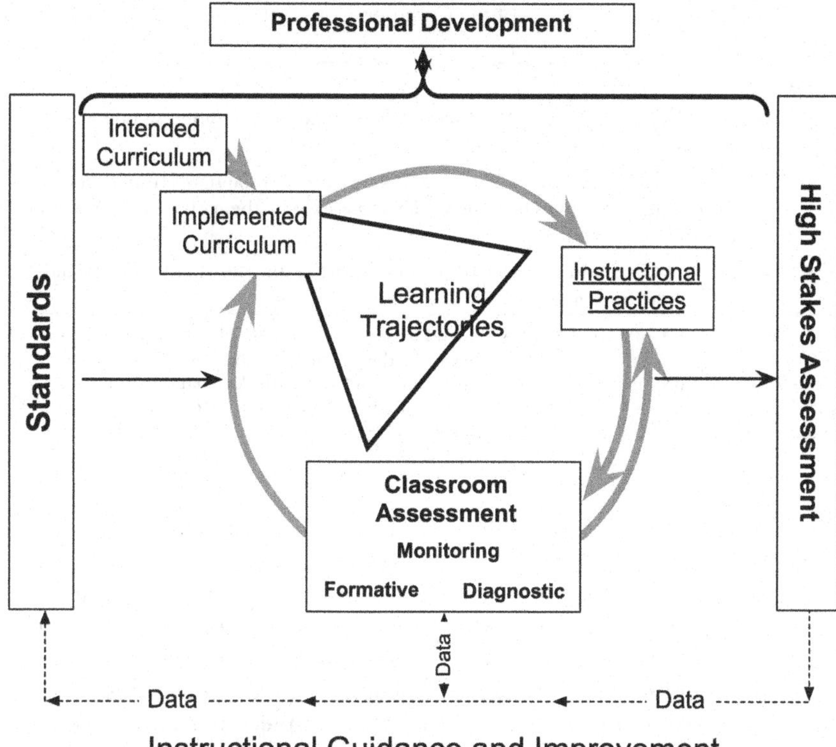

FIGURE 5.1. Learning trajectories within the instructional core and accountability system. *Note:* Learning trajectories as referents for coordination and translation among standards, assessment, curriculum, and professional development (Confrey, 2012).

for Best Practices & Council of Chief State School Officers, 2010, p. 5). The document incorporates research in student learning inasmuch as the accelerated writing schedule could accommodate.

Despite the writers' intent however, learning progressions are difficult to discern in the standards document, due to a combination of the grade-by-grade listing, the hierarchical organization, and varying grain sizes among different standards. The CCSS-M document (typical for most standards documents to date) does not readily indicate the relationships of topics and concepts within any single year of schooling, nor how to readily trace how one idea or concept leads to another over time. The CCSS-M document is organized hierarchically, as shown in Table 5.1.

TABLE 5.1. Hierarchical Structure of the Common Core State Standards for Mathematics

[1.] Grade
[2.] Domains: larger groups of related standards. Standards from different domains may sometimes be closely related
[3.] Clusters of groups of related standards. Note that standards from different clusters may sometimes be closely related, because mathematics is a connected subject.
[4.] Individual standards that define what students should understand and be able to do

Note. Excerpted from Common Core State Standards for Mathematics, by the National Governors Association Center for Best Practices and the Council of Chief State School Officers, 2010, Washington, DC. Retrieved from National Governors Association Center for Best Practices and the Council of Chief State School Officers website for Common Core State Standards Initiative: http://www.corestandards.org/assets/CCSSI_Math Standards.pdf

The numbering system in the document makes it challenging to readily understand how one particular standard in one year proceeds to those that most reasonably follow next year in a progression (for example, Standard 1 in a cluster in grade 3 may not readily follow, instructionally, Standard 1 in the same cluster in grade 2). In its published form, the standards document is not a convenient tool for instructional planning or curricular design and evaluation. In standards, as in mathematics, multiple representations offer many advantages. Most users need some other way to engage with the organization and content of standards in order to readily understand them and to design their instruction to ensure coherent coverage of content described in the standards.

How then to transform the standards into coherent progressions of topics, and to bring such a complex body of information more to life for teaching, professional development, and curriculum and assessment development? Learning trajectories aim specifically to delineate the cognitive progression of concepts and ideas from one to the other, over time. A representation that arranges and displays the standards to embody an explicit framework of learning progressions within and across grade levels could transform a standards document into a user-friendly tool that supports instruction, curriculum development, assessment development, research in student learning, and even student reflection about their own learning. Such a display could form the basis for phased introduction of targeted professional development to the CCSS-M for both mathematical content and instructional practice and could serve as a pointer for ongoing research into learning progressions and future revisions of the standards themselves.

Our aim in developing a novel learning trajectories-based display of standards has been to create *extensible* tools for both print and electronic displays that could accommodate an increasing diversity and volume of support material reflecting research into student learning. Our first foray into such an endeavor was a simple two-dimensional chart that became a framework for developing the North Carolina state standards in 2009 (prior to the development of the CCSS-M). We have now developed two different displays of the CCSS-M in which we have represented the standards to reflect our best conjectures of learning trajectories across the various concepts covered in K–12 school mathematics. These representations derive from the earlier work with the North Carolina standards, our research group's expertise, and our understanding of other investigators' research on learning. These displays lay the groundwork for more detailed articulation of learning trajectories and, we anticipate, an eventual framework for revising and updating the CCSS-M in the coming years. Both representations are built for the most common display format in most classrooms, that is, a flat surface such as a wall or a flat-screen monitor.

DESIGN PRINCIPLES FOR LEARNING TRAJECTORY-BASED DISPLAYS OF MATHEMATICS STANDARDS

Several design principles for representing learning trajectories, established during development of the 2009 North Carolina standards and refined in the context of the CCSS-M, guided the development of these Common Core standards displays. They emphasize the displays' potential to serve as boundary objects for the various communities within mathematics education.

1. To improve coherence of an educational system, teachers, students, parents, and administrators must have a common, intelligible framework for predicting, monitoring, and understanding students' progress over time in (a) understanding increasingly complex mathematical concepts, (b) reasoning about mathematical topics and problem solving, and (c) skills that students are expected to learn from year to year. Readily interpretable visual displays of this framework are crucial for all these stakeholders in children's education.
2. The learning of mathematics—students' development of mathematical understanding and reasoning—is *not* synonymous with the logical structure of the mathematical disciplines. The former is grounded in activity and in the progressive learning and iterative reorganization of concepts that arise through instruction, classroom discourse, and student work, and their interplay. By contrast, the "logic of the discipline" is the product of hundreds of years of research in mathematics, statistics, and engineering, and is aptly captured in the formalizations of definitions, theorems and proofs,

compact and concise notation in sequencing of concepts based on logical decomposition (often retrospectively constructed) of ideas, and in the application of mathematics to solve interdisciplinary problems. While it informs the main goal understandings of learning trajectories, it does not generally represent or embody the conceptual development experienced by the vast majority of children.

3. Learning trajectories based on the experience and judgment of experts in student mathematics learning should be embedded within the major topic strands of curriculum standards (for example, number and operations, measurement, probability and data/statistics, geometry, algebra, and for high school, discrete mathematics). As further empirical research on learning trajectories/progressions is conducted, the findings from this research should inform revisions of both the conjectured learning trajectories and the applicable standards.
4. Representations that display the progression of expected student learning from year to year with respect to the standards illustrate the relationship of topics to each other and/or integrate the content strands within each grade level could serve as a valuable common reference tool for multiple communities in education. Learning trajectory-based displays of standards focus attention on concepts and skills at an individual grade level and on continuity of conceptual development from year to year. The latter is of particular importance when, as is the case with CCSS-M, new standards are more demanding of instruction and learning and when there is more material per year and less time for review and relearning (teachers must prepare their students to learn the current year's material without spending instructional time repeating previous year's material). Clear displays of expected student progression will assist in professional development planning as well.

LEARNING TRAJECTORY CHARTS OF THE COMMON CORE STANDARDS[2]

The first of these representations is modeled on one we created during 2009, when we served as members of the team revising the state mathematics standards for the state of North Carolina (before CCSSO and NGA began writing the CCSS). The North Carolina standards writing team, at the suggestion of author Confrey, endeavored from the outset to construct and describe standards in terms of progression of concepts and skills that students would develop over time, based on research in student learning. To facilitate this, the North Carolina writing team agreed to use a visual representation as a framework in which to craft the standards so that all proposed standards were (a) shown by progression of topics in each content strand from year to year, and simultaneously (b) aligned within each year to make it more obvious (visually) whether they were all reasonable for students to accomplish in that particular year. The result was a two-dimensional chart display, with time denoted left to right (grade level, for K–8, and level of complexity for 9–12) along the horizontal axis, and content strands and substrands—learning trajectory titles—listed along the vertical axis. This meant, first of all, that all standards for each grade (for grades K through 8), or those for corresponding topic difficulty (for high school), were all listed in a single column: the standards revision task force, teachers, students, and parents could readily see what content should be accomplished in any given year. Most important, however, was that standards for each content strand were listed horizontally, (proceeding across time, that is, temporally from left to right, in the case of K–8 standards, and from lower to higher complexity in the case of high school standards). This meant each particular content strand or substrand is displayed as a progression of content from less sophisticated to more sophisticated along the time axis, corresponding to the revision team's best judgment from student learning research and from teacher experience. It also meant that all teachers can readily discern what they are responsible for teaching in each year, and further, what preparation to expect from students coming into their class at the beginning of the year, and what their classes must accomplish by the time they proceed to the next year's classes. The overall effect is that the standards were designed, from the outset, in a coherent framework sensitive to student conceptual development.

The spreadsheet display was not treated as a set of learning trajectories because the grain size of the standards was far too large. However, as the North Carolina writers used this spreadsheet display as a tool for developing the state standards, they were able to (1) ensure that the standards made sense as content and cognitive progressions and (2) align content standards for coherence within any given year of school. The writing team applied the best available knowledge in the field from both research and practice perspectives (the writing team included master teachers, curriculum and professional development experts, and mathematics education researchers). Use of the display as a framework facilitated the identification of gaps in progression of topics, improved consistency of the grain size of the standards overall, and supported the team's ability to efficiently focus on and resolve disagreements on topic inclusion and sequencing, or to reach appropriate compromises—that is, to make shared decisions.

CCSS-M Standards Charts for Grades K–8

These displays of NC standards were made available to the CCSS-M writing team to support the development of the new Common Core standards.

TABLE 5.2. Organization of Content Strands for CCSS-M Standards Chart for Grades K–5

Content Strand \ Grade	K	1	2	3	4	5
Quantity, Measurement, and Data						
Length						
Units and Conversion						
Area, Perimeter, and Volume						
Money and Time						
Data and Statistics						
Numeration, Operations, and Algebraic Thinking						
Counting						
Place Value and Decimals						
Equipartitioning and Naming						
Fractions and the Number Line						
Fraction Equivalence						
Comparison of Fractions						
Concepts, Strategies, and Properties of Addition and Subtraction of Whole Numbers, Fractions, and Decimals						
Regrouping of Addition and Subtraction of Whole Numbers						
Concepts, Strategies, and Properties of Multiplication and Division of Whole Numbers, Fractions, and Decimals						
Word Problems in Any of the Four Operations						
Fluency in All Four Operations						
Evaluating Expressions or Equations						
Solving for Unknowns; Coding Verbal Expressions						
Algebraic Patterns						
Geometry						
Naming, Identifying Attributes, and Defining						
Composing or Decomposing						
Distinguishing Dimensions						
Angles and Coordinate Graphing						

Immediately after the CCSS-M document was released, we developed the "Learning Trajectory Display of the Common Core State Standards for Mathematics," or "standards charts," applying the same design principles used in developing the North Carolina standards. We created three standards charts, one for grades K–5, one for grades 6–8, and one for high school. We partnered with Wireless Generation to distribute the charts as printed posters (www.wirelessgeneration.com/posters).[3]

Tables 5.2 and 5.3 illustrate the overall strand and substrand organization of the standards charts for grades K–5 and 6–8, respectively. In the standards charts, each Common Core standard is reproduced verbatim. Each block of cell text comprises an individual standard (or subpart of a standard), along with that standard's alpha-numeric code: chart users can readily cross-reference the chart with the CCSS-M document. The CCSS-M vary considerably in grain size; for multipart CCSS-M standards (containing

TABLE 5.3. Organization of Content Strands for CCSS-M Standards Chart for Grades 6–8

Content Strand / Grade	6	7	8
Ratio and Proportional Relationships and Percent			
Ratio, Rate, and Slope			
Word Problems			
Rational Number System and Operations and Introduction to Irrationals			
Whole Numbers, Rationals, and Irrationals			
Negative Numbers			
Locating and Operating with Rational Numbers in 1D and 2D space			
Word Problems and Rational Numbers			
Algebraic Reasoning			
Exponents, Roots, and Scientific Notation			
Expressions			
One Variable Equations, Inequalities, and Word Problems			
Simultaneous Linear Equations			
Introduction to Functions and Linear Functions			
Geometry			
Area and Volume			
Angles, Coordinate Plane, and Transformations			
Similarity and Congruence			
Pythagorean Theorem			
Probability and Statistics			
Statistical Investigations and Sampling			
Descriptive Statistics (Central Tendency and Distribution)			
Bivariate Data and Scatter Plots			
Probability			

parts a, b, and c, for instance), the parts may be separated and located in different trajectories (rows). Main topic domain names reflect those identified in the CCSS-M document; many of these were condensed or combined, especially when the domain names in the CCSS-M vary across the years of school. Individual content strands were named to reflect research-based learning trajectories where possible, content strands from other sets

TABLE 5.4. One Sample of Standards from the Learning Trajectories Display of the CCSS-M, Grades 6-8: Ratio, Rate, and Slope, Within the Ratio and Proportional Relationships and Percent Cluster

Content Strand	Grade 6	Grade 7	Grade 8
	Ratio and Proportional Relationships and Percent		
Ratio, Rate, and Slope	Understand the concept of a ratio and use ratio language to describe a ratio relationship between two quantities. For example, "The ratio of wings to beaks in the bird house at the zoo was 2:1, because for every 2 wings there was 1 beak." "For every vote candidate A received, candidate C received nearly three votes." [6.RP.1]	Compute unit rates associated with ratios of fractions, including ratios of lengths, areas and other quantities measured in like or different units. For example, if a person walks 1/2 mile in each 1/4 hour, compute the unit rate as the complex fraction (1/2)/(1/4) miles per hour, equivalently 2 miles per hour. [7.RP.1]	Graph proportional relationships, interpreting the unit rate as the slope of the graph. Compare two different proportional relationships represented in different ways. For example, compare a distance-time graph to a distance-time equation to determine which of two moving objects has greater speed. [8.EE.5]
	Understand the concept of a unit rate a/b associated with a ratio a:b with b ≠ 0, and use rate language in the context of a ratio relationship. For example, "This recipe has a ratio of 3 cups of flour to 4 cups of sugar, so there is 3/4 cup of flour for each cup of sugar." "We paid $75 for 15 hamburgers, which is a rate of $5 per hamburger." [6.RP.2]	Recognize and represent proportional relationships between quantities. a. Decide whether two quantities are in a proportional relationship, e.g., by testing for equivalent ratios in a table or graphing on a coordinate plane and observing whether the graph is a straight line through the origin. d. Explain what a point (x, y) on the graph of a proportional relationship means in terms of the situation, with special attention to the points (0, 0) and (1, r) where r is the unit rate. [7.RP.2.ad.i]	Use similar triangles to explain why the slope m is the same between any two distinct points on a non-vertical line in the coordinate plane; derive the equation y = mx for a line through the origin and the equation y = mx + b for a line intercepting the vertical axis at b. [8.EE.6]

of standards (for instance from the most recent version of the North Carolina state standards), or commonly used names from curricula; the CCSS-M cluster names were approximated when possible.

Though the charts are too large to permit their legible reproduction in book format, we have presented a sample of one strand, that for ratio, rate, and slope (a substrand of the ratio and proportional relationships and percent strand), in Table 5.4. This sample illustrates most of the features described above.

CCSS-M Standards Chart for High School: Mediating the Dual Curriculum-Dual Standards Dilemma

High School Standards in North Carolina

In 2009, the writing group for the North Carolina standards (of which the authors were members) made a principled decision that there should be only one set of mathematics standards for the state, regardless which curricula were used. This was a major innovation in state mathematics standards. Previously, as third-party organizations (Achieve, College Board) and some other states (Missouri, Minnesota) had done, North Carolina had been developing course-specific high school standards and was planning to develop separate high-stakes end-of-course exams for the traditional and integrated mathematics courses. This implied that integrated and traditional mathematics curricula would have different mathematics standards. The North Carolina standards writing team recognized multiple practical and conceptual downsides pertaining to maintaining dual standards:

- Multiple sets of standards and multiple exams required duplication of effort and expense in development.
- Distinct standards for integrated mathematics could lead to inadvertent bias toward a particular curriculum, undermining subsequent district or school flexibility to adopt improved integrated mathematics curricula later.
- Different sets of standards implied undesirable duplication of professional development efforts and expense.
- Perhaps most important, for a state to have two different sets of standards gave the appearance of inconsistency or even disequity for students taking different curriculum types. This led to the impression that students would learn "different" mathematics depending on their curriculum.

The North Carolina standards writers, after considerable discussion, asserted that the state should have only one set of standards for all students. The writers argued that a single set of standards would make it clear, regardless of curriculum, that by the end of the first three years of high school (the period covered by mandatory state standards), all students should have learned the same mathematics content. This decision was based in part on equity considerations, as well as considerations of the expense of parallel end-of-course assessments. The writing team therefore extended the learning trajectories approach into high school mathematics. This decision permitted the standards writing team to focus on progressions of mathematical content learning for the high school standards instead of de facto course outlines. The standards chart therefore also became the format for developing the high school standards, organized around large content domains (algebraic reasoning, functions, probability and data analysis, geometry, discrete mathematics, and mathematical modeling). The horizontal (time) direction of the high school standards was denoted in levels of increasing difficulty or complexity and sophistication of reasoning, instead of grade level. Vertical alignment of groups of standards was then done on the basis of content coherence among various strands (similar to the alignment of standards within grades for K–8). So when viewed across all the strands simultaneously from left to right, the standards reflected an integrated mathematics module, but when viewed along a cluster of like strands (algebra, for instance) the standards reflected, in the main, the siloed courses.

For North Carolina, the next question to be resolved was how to assess students' accomplishment of a single set of standards while allowing flexible district or school curriculum choices. Establishing coherence among standards and assessment had been made a top priority, as set out by the North Carolina Blue Ribbon Commission on Testing and Accountability (2008). The situation was resolved by determining to assess a basic set of standards after the first year of high school study, to assure a solid foundation for the remaining two years of high school mathematics regardless of curriculum. This would require some adjustments by teachers of both curriculum types. The standards covered in the first-year test would include much of traditional first-year algebra, along with some geometry and a substantial amount of probability and statistics. There would be no statewide testing after the second year, reflecting the diversity in curricula. At the end of three years, however, a statewide assessment would cover all the content required by the second- and third-year standards. The result of this overall strategy was to generate a consistent set of statewide standards for all students; flexibility in the choice of curriculum by teachers, schools, and/or districts; a reduced load of high-stakes assessment for the students; and forecasted higher efficiency and lower costs for the state. The standards and the new assessment strategy were adopted by the State Board of Education in 2009.

The CCSS-M High School Standards

The following year, CCSS-M mirrored this principle—that there should be a single set of mathematics learning standards, including high school, and that all students must learn the same mathematical concepts and skills, regardless of what particular curriculum is deployed in a particular school or district. The CCSS-M standards document went farther, in fact, specifically asserting that the CCSS-M is not intended to specify any curriculum or pedagogical approach:

> These Standards do not dictate curriculum or teaching methods. For example, just because topic A appears before topic B in the standards for a given grade, it does not necessarily mean that topic A must be taught before topic B. A teacher might prefer to teach topic B before topic A, or might choose to highlight connections by teaching topic A and topic B at the same time. Or, a teacher might prefer to teach a topic of his or her own choosing that leads, as a byproduct, to students reaching the standards for topics A and B. (National Governors Association Center for Best Practices & Council of Chief State School Officers, 2010, p. 5)

Seeley suggested that it remained to be seen whether the CCSS-M would support the incorporation of "topics not traditionally addressed in algebra and geometry courses (e.g., statistics, linear programming, mathematical modeling) [that] are increasingly recognized as critical in our rapidly changing world, and various forms of technology [that] have become crucial as tools and as objects of study" (Seeley, 2008, p. 3). In their final form, they explicitly support much deeper statistical reasoning, beginning in 6th grade; they support a level of mathematical modeling; little or no linear programming content is implied. Nonetheless, by explicitly supporting flexibility in curriculum selection, and rejecting the primacy of one curriculum over another, CCSS-M has made an important step toward reversing what Seeley had observed, namely that "high school mathematics in the United States is stuck" (2008, p. 1).

The CCSS-M, like the North Carolina 2009 standards, left unspecified any grade levels for particular high school mathematics topics. We note that K–8 mathematics is essentially an integrated approach to mathematics, even though it delineates content by school year. To represent the high school standards, absent specification of grade levels, our approach was essentially the same as it had been for the North Carolina standards: to (1) lay out content strands in sequences that reflect increasing complexity of the topic, using the horizontal (time) axis as a proxy for complexity, and (2) align standards from different trajectories vertically so that the difficulty of standards in different strand in same column are mutually consistent.

The general organization of the high school standards chart is shown in Table 5.5. Instead of specifying grades, successive proficiency levels are indi-

TABLE 5.5. Organization of Content Strands for CCSS-M Standards Chart for High School

Content Strand / Level	1	2	3	4	5	6	7	8
Number and Quantity								
Exponents, Radicals, Complex Numbers, and the Fundamental Theorem of Algebra								
Units and Quantities								
Vectors and Matrices								
Algebra								
Definitions and Proofs								
Expressions and Formulas								
Graphing								
Linear, Polynomial, and Rational Functions								
Exponential Functions								
Systems of Equations and Inequalities								
Functions								
Function Definition, Operations, Composition, and Inverse								
Sequences								
Rate of Change and Characteristics of Graphs								
Quadratic Functions								
Trigonometric Functions								
Transformations								
Modeling with Functions								
Geometry								
Proofs of Theorems								
Triangle Congruence								
Similarity and Sectors								
Trigonometry								
Parallel and Perpendicular Lines and Parallelograms								
Constructions and Circles								
Transformations and Analytic Geometry								
Conic Sections								
Area and Volume								
Modeling								
Statistics and Probability								
Sampling and Design								
Data Distributions								
Probability								
Bivariate Data, Linear Regression, and Correlation								

cated for each strand and substrand. Up to eight levels were constructed for some topics. Space constraints prevent us from including an entire chart in this volume. As before, the general layout of the charts can be viewed at (http://www.wirelessgeneration.com/posters).

HEXAGON MAP OF THE COMMON CORE STANDARDS

The second representation we developed for the Common Core standards is a "hexagon" map of the Common Core standards (Confrey, Nguyen, Lee, Panorkou, Corley, & Maloney, 2011). The initial version of this form of standards map was developed by author Confrey (in collaboration with Wireless Generation) to depict the New York City elementary mathematics standards in 2006. As suggested by the name, each standard is assigned to an individual hexagon, labeled with an abbreviated version of the text of the standard. Both time and mathematical complexity are indicated in the hexagon map as flow from lower left to upper right. Learning trajectories are represented as swaths of hexagons of one color, progressing along a more or less diagonal path from lower left to upper right to represent conceptual and skill growth across time. Grade levels are indicated with hexagons in bands of the same background that cut approximately at right angles across the learning trajectory diagonals. The text of each standard is further color-coded to signal the CCSS-M cluster (main content strand) of the standard. On the turnonccmath.net website, passing the cursor over a hexagon will cause highlighting of either the entire grade level or the learning trajectory of which the particular hexagon/standard is a member.

A version of the K–8 hexagon map color-coded by grade levels is shown in Figure 5.2. In this version, the hexagon background colors correspond to the grade level of the standards (K–8). A second version of the hexagon map identifies the main content strands and learning progressions within the standards (Figure 5.3). We also refer the reader to www.turnonccmath.net for the two versions of the hexagon display of K–8 CCSS-M standards (grade level and learning trajectory).

The grain size of the CCSS-M standards varies considerably. Some topics in the CCSS-M also vary in degree of continuity across years. The standards charts, a tabular display, reflected our best conjectures of standards continuity and of the school years in which standards occur (or don't occur) for each learning trajectory. However, this resulted in many spans of empty cells that appear to leave large gaps in conceptual continuity of mathematics learning. In the hexagon maps, the hexagons are set adjacent to each other to represent learning progressions discerned in and developed from the Common Core standards. This closer packing visually supports juxtaposition of topics within and between progressions. Juxtaposition of related content is at the same time constrained, of course, by the two-dimensional layout of the hexagons. Some strands had to be placed at considerable dis-

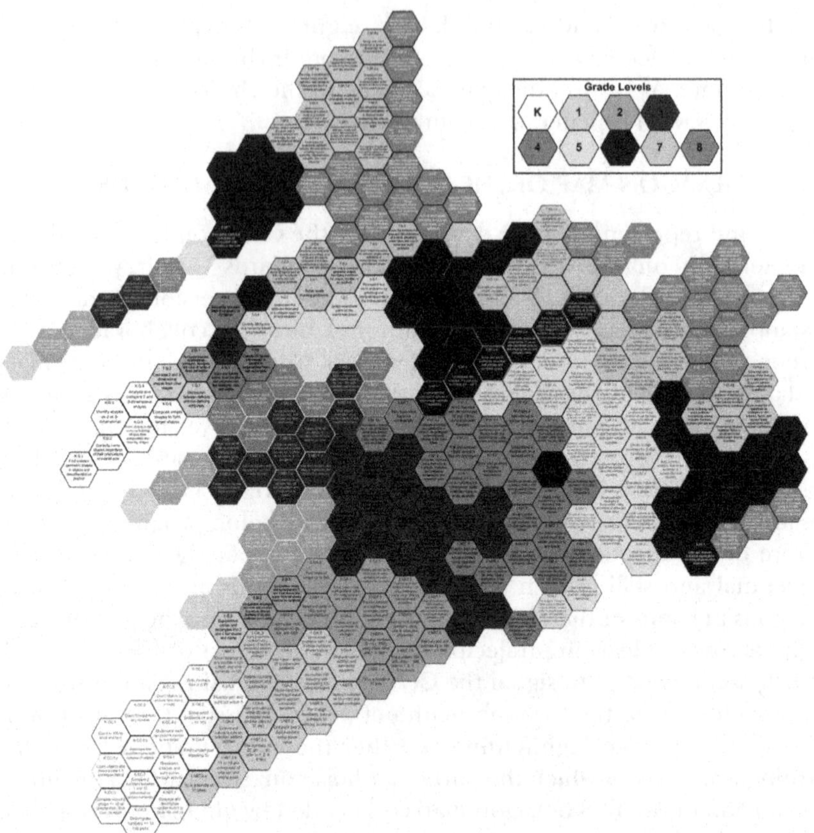

FIGURE 5.2. Hexagon map of the Common Core State Standards, K–8. *Note:* Hexagon map of the K–8 Common Core State Standards for Mathematics, with grade levels indicated by background colors of hexagons, shown here in shades of gray. (See also http://www.turnonccmath.net for a color version that can be enlarged.)

tance from each other, even if they are conceptually related. (For instance, statistics and probability is placed at some distance from rational number topics, even though probability and rational number have many commonalities.)

Learning Trajectories in the Hexagon Map Format

The learning trajectories depicted in the hexagon map in Figure 5.3 represent a major refinement of the CCSS-M into a series of learning trajectories. These learning trajectories are developed in detail ("unpacked")

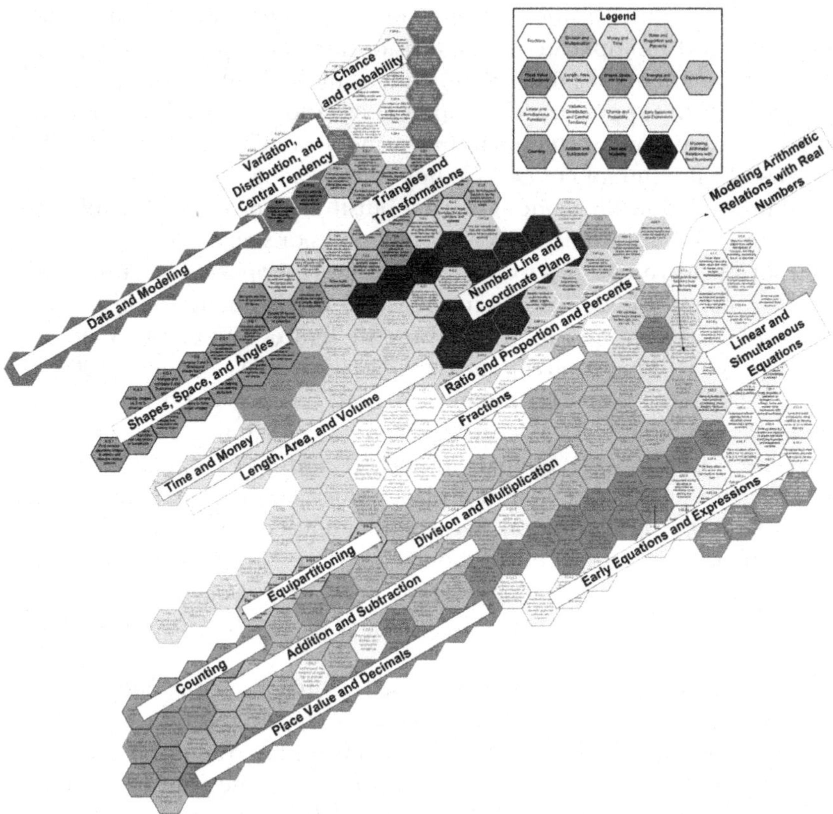

FIGURE 5.3. Learning trajectories in CCSS-M hexagon map. *Note:* Hexagon map of K–8 CCSS-M, with main content strand learning trajectories color-coded, and overlaid with name labels for each trajectory, shown here in shades of gray. (See also http://www.turnonccmath.net for a color version that can be enlarged.)

via extensive textual *descriptors*, including additional bridging standards, at turnonccmath.net. The directionality of the learning progressions is generally from lower left (more fundamental and basic) to upper right (more advanced, sophisticated, and complex). The learning trajectory names, labeled in Figure 5.3, emphasize that the fundamental domains, in the lower grades, are interrelated and comprise the foundation for higher-order mathematical reasoning in later grades. Close juxtaposition of the domains in the lower left of the map reflects how the strands of basic mathematical concepts are required for student proficiencies to build into more complex strands. For example, (1) counting, addition and subtraction, equipartitioning, and place value all are needed as foundation for the development of proficiency in multiplication and division; and (2) measurement, equipartitioning, and multiplication and division are three progressions that most directly combine to generate the understanding and use of fractions. These also in turn support the development of ratio and proportional reasoning and rational and irrational number understanding. Two-dimensional displays of this kind, of course, are not entirely adequate to map all the possible interrelationships among topics, but this kind of visual mapping provides a major improvement over a list of topics or standards.

Many of the learning trajectories are extensively supported by empirical evidence, much of which is noted in the CCSS-M document: for instance, composition and decomposition of geometric figures (Clements, Wilson, & Sarama, 2004); length, area, and volume (Barrett & Clements, 2003; Battista, 1999, 2007; Battista, Clements, Arnoff, Battista, & Borrow, 1998; Clements & Sarama, 2009); counting (Clements & Sarama, 1992; Fuson, 1988; Steffe & Cobb, 1988); equipartitioning (Confrey et al., 2009; see also Chapters 1 through 4, this volume); place value (Fuson, 1992; Ma, 1999; Thompson, 1999); and addition and subtraction (Carpenter, Franke, Jacobs, Fennema, & Empson, 1997; Fuson et al., 1997; Ma, 1999). Others are less fully substantiated: Less directly relevant research has been conducted on student learning on other specific topics (e.g., functions, expressions, and equations).

The hexagon map and the accompanying learning trajectory descriptors therefore represent a combination of research results, professional judgment, and teaching experience that have contributed to multiple rounds of standards writing and revisions. They represent a large synthesis of the accumulated experience of multiple teams of educators and researchers, including the authors of turnonccmath.net; research and advice of many other researchers; the expert teachers, professional development experts, and state education department curriculum personnel who developed the North Carolina Mathematics Standards; author Confrey's experience with the CCSS-M writing and validation process; and the authors' experience (along with that of co-editor Nguyen) in representing standards in various formats. The visual model presented here comprises a theory of learning for the mathematical topics included in the standards. Other visual models for the CCSS-M are possible, of course, and we anticipate that this work will promote alternative representations that would be useful in different contexts.

Abridged and Unabridged Learning Trajectories in the Hexagon Map Format

All the learning progressions represented in the hexagon map itself and in the standards charts must be regarded as abridged, or condensed, rather than unabridged, or complete learning trajectories, because they incorporate only the Common Core standards, which themselves represent a curtailed version of mathematics content or student learning. How then to build on the CCSS-based learning trajectories and construct more "unabridged" learning trajectories, with a more consistent grain size, which better represent intermediate states of learning, and more adequately support instruction and planning in classrooms, professional development programs, and curriculum or assessment development? We illustrate this with two examples: ratio and proportion and percents trajectory, and the equipartitioning trajectory.

Equipartitioning Learning Trajectory

A learning trajectory for equipartitioning has been developed by the DELTA project (see Chapter 3, this volume; and Confrey et al., 2009.) Several components of the equipartitioning learning trajectory are identified in the CCSS-M standards; these standards' approximate orientation to each other within the K–12 hexagon map (Figure 5.3) is displayed in Figure 5.4a. In building an unabridged learning trajectory in hexagon format, we first rearranged the existing Common Core standards to align with the equipartitioning learning trajectory described in Chapter 3 (Figure 5.4b). We then added many of the proficiency levels, or *bridging standards* (Figure 5.4c), so named because they serve to bridge from the existing Common Core standards to an unabridged learning trajectory. The structure of the hexagon layout helps emphasize that an instructional pathway is not dictated and is not necessarily linear. The juxtaposition of standards (CCSS-M and bridging) depicts cognitively related proficiency levels in each learning trajectory but is meant to leave the particular instructional path to the instructor's choice or a curriculum writer's best judgment.

Ratio and Proportion and Percent Learning Trajectory

A learning trajectory for ratio and proportional reasoning and percents was developed for the Grades 6–8 CCSS-M. Figure 5.5a displays the relevant section of the K–12 hexagon map and maintains the approximate grade topic ordering recommended in CCSS-M. The CCSS-M ratio and proportion standards are of a relatively consistent level of conceptual detail (grain size). The trajectory as we have defined it includes two eighth-grade standards from the expressions and equations (EE) standards, 8.EE.5 (graphing and comparing proportional relationships, and interpreting unit rate as slope) and 8.EE.6 (explaining slope with similar triangles and deriving equations for lines), because of their critical role in representing and comparing proportion and ratio relations. The seventh-grade ratio and proportion (RP) standard 7.RP.2d is not contiguous on the hexagon map with the other standards in the same trajectory; this standard emphasizes finding

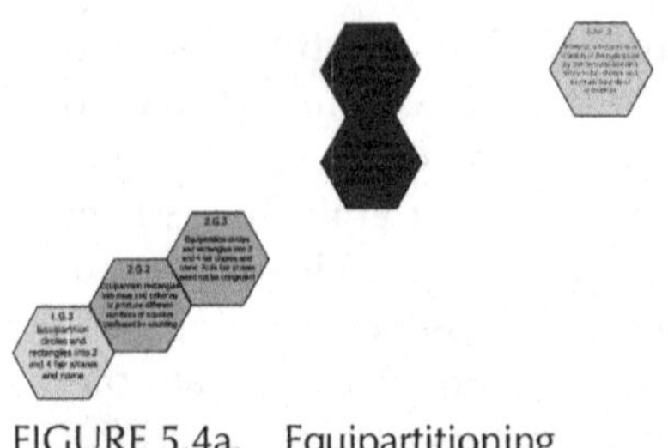

FIGURE 5.4a. Equipartitioning learning trajectory as identified in the CCSS-M hexagon map.

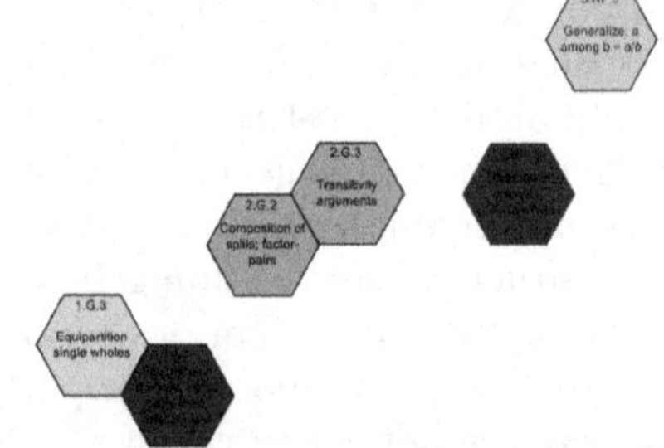

FIGURE 5.4b. Equipartitioning standards rearranged to match equipartitioning learning trajectory matrix (per Confrey et al. in Chapter 3, this volume).

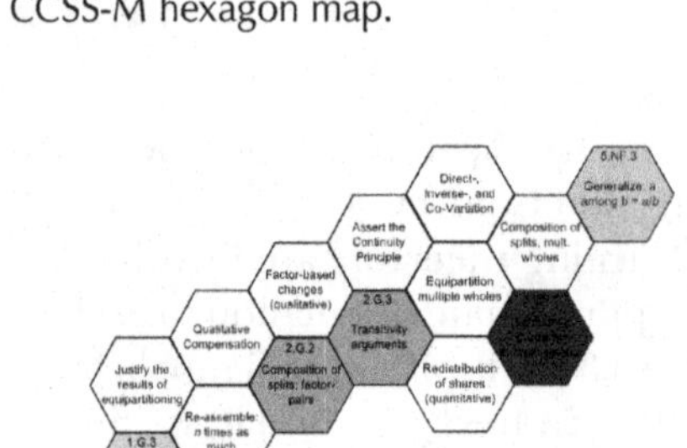

FIGURE 5.4c. Equipartitioning standards from FIGURE 5.4b with bridging standards added. (See also http://gismo.fi.edu/Chapter5Fig4 for electronic versions that can be enlarged.)

FIGURE 5.4. Hexagon map sequence of equipartitioning learning trajectory.

coordinate points, and in particular, finding points on lines that represent unit rates in proportional relations, so the standard was placed nearer standards in the integers, numbers lines, and coordinate planes, trajectory in the K–12 map.

Our experience in studying ratio reasoning (Confrey & Scarano, 1995; Scarano & Confrey, 1996), including discussions during revisions of the North Carolina standards, led us to conjecture an "unabridged" ratio and proportion learning trajectory that included bridging standards. Standards 6.RP.1 and 6.RP.2 subsumed an enormous amount of fundamental ratio understanding (large "grain size"), so we added four bridging standards around those two Standards (unshaded hexagons 6.RP.A through 6.RP.D; Fig. 5b). These highlight the importance of textured, gradual development of student understanding of the ratio concept, which in turn supports deeper rational number reasoning. The bridging standards specify the use of simple tables (which we have termed *ratio boxes*), graphs, and covariation and correspondence to focus on the multiplicative relationships within ratios, and to identify—and distinguish between—unit ratio (in the form $1/a$ or $a/1$) and ratio unit (or base ratio, in the form a/b). The latter provides important conceptual foundation for more general development of proportional reasoning (unit rates and ratios in the form of rational numbers) in subsequent Standards. The proficiencies represented in these four bridging standards can be readily incorporated in a Grade 6 instructional plan.

Standards 7.RP.2 a through d focus on investigation and identification of proportional relationships with multiple representations, including equations of the form $y = kx$. These standards incorporate representation of ratios as unit rates and as constants of proportionality with tables, graphs, diagrams, and equations, and therefore include moves to relate the symbolic (equation) treatment of direct variation with other representations of ratio contexts (including straight lines on a coordinate plane). The specific arrangement of the ratio and proportion standards (Fig. 5.5a and 5.5b) reflects our view that students will more readily master the content of Standards 7.RP.2.a through d if instruction interweaves development and use of representations with real world situations requiring mathematical modeling with ratio (6.RP.3.a through d). Note that this ordering of standards with the LT does NOT dictate the assessment plan for end-of-grade tests, but does emphasize the instructional experience of developing the ratio concept using multiple representations linked to real world problems to strengthen students' understanding of both the ratio concept and its relevance. Finally, use of ratios of fractions (7.RP.1) becomes, in this LT, an extension of that robust ratio reasoning, not a precursor to (or a potential distraction from) conceptual mastery.

Bridging standard 6.RP.E explicitly highlights the concept of ratio ($a{:}b$ or a/b) as an operator. Ratio is used as an operator not only in unit and measurement conversion problems, but across proportional reasoning situations, and then later is central in understanding linear functions, so a bridging standard focusing on that concept is appropriate. Bridging standard 7.RP.A emphasizes the importance of students legitimizing the relationship $a/b = c/d$ if $ad = cb$ for themselves as a culmination of developing the ratio concept, rather than merely adopting cross multiplication as a procedure without understanding.

Standard 8.EE.5 focuses on comparing ratios, and interpreting the concept of slope in graphs of points arising from different ratio relationships, and should be well integrated with the seventh grade ratio and proportion standards; it could even be legitimately placed prior to or simultaneous with instructional treatment of those two seventh grade standards. Standard 8.EE.6 deals explicitly with a geometric approach to demonstrating the constancy of slope for any pair of points on a line, and distinguishing direct variation from nonproportional variation through the explicit derivation of the equations $y = kx$ and $y = mx + b$. Bridging standard 8.EE.B was included in this learning trajectory, to highlight the importance of students' understanding negative slope, not just positive slope.

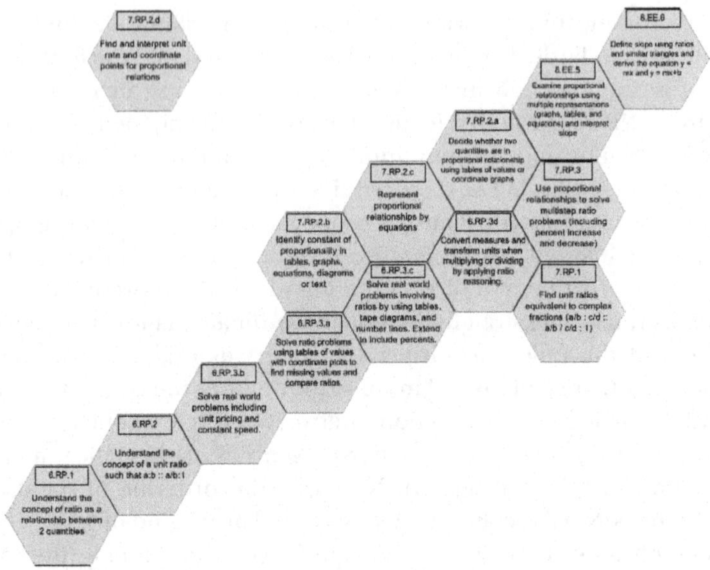

FIGURE 5.5a. Learning trajectory for ratio and proportion and percents, from CCSS-M hexagon map.

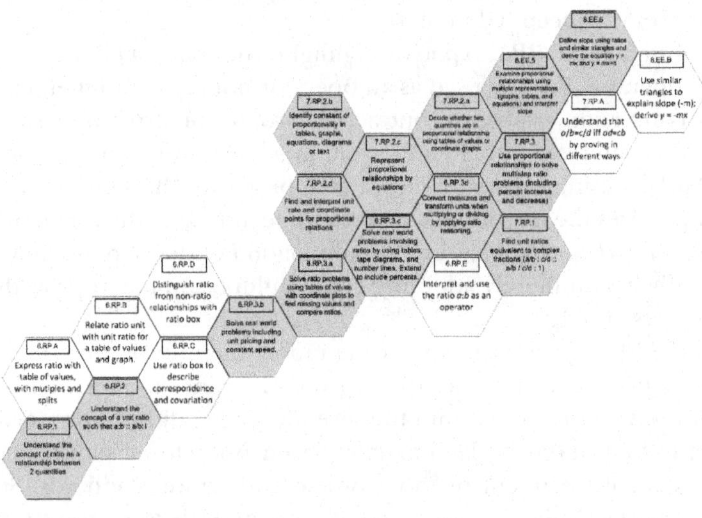

FIGURE 5.5b. Ratio and proportion and percents learning trajectory, rearranged and with bridging standards added. (See also http://gismo.fi.edu/Chapter5Fig5 for electronic versions that can be enlarged.)

DISCUSSION AND SUMMARY

Standards are a negotiated policy document that must be interpreted by the users of that document. Those users include teachers, teacher educators, administrators, parents, and, we would argue, students. Learning trajectories are based on an inherent notion of development of student learning over time. Learning trajectories indexed to the new Common Core State Standards for Mathematics can comprise a set of boundary objects for framing instructional planning and expectations and necessary professional development far more flexibly than is possible with mere textual lists of standards. They can comprise a new tool for implementing the standards in classrooms and for discussing, critiquing, adapting, and eventually revising them. The learning trajectory displays of the standards presented in this chapter can fulfill key roles of boundary objects: supporting cooperation *among* different communities while facilitating greater specification and utility *within* each community as needed.

The Common Core State Standards are a major component of a nationwide effort to generate coherence in mathematics instruction and learning across educational systems and states and to deepen American students' mathematical understanding and readiness for higher education and careers in the 21st century. These displays, especially the hexagon map and the materials provided at turnonccmath.net, embed the standards within empirically supported networks of constructs that students traverse to accomplish increasingly sophisticated or complex levels of mathematical reasoning. The overall aim of these displays is to make it easier for educators and others to interpret the CCSS-M in terms of student learning.

Organizing the standards into learning trajectories in a graphical display and communicating the development of student learning over time will, we hope, make the CCSS-M more readily useful to teachers, teacher educators, and other communities within education. They can support teachers in planning and focusing their instruction within a single grade level, as well as cross-grade instructional collaboration to ensure consistent instruction and student conceptual growth year to year. The learning trajectories, including the descriptors, assist teachers in recognizing where students are struggling and in developing targeted instruction to mitigate against students disengaging and falling behind. The learning trajectory displays and descriptors also support teachers in deepening their own content knowledge, as we have recently verified in a professional development class for in-service K–7 teachers. We believe that in the larger educational community, the learning trajectory charts, maps, and descriptors can help focus discussion of student learning and provide a basis for refining the underlying standards and their representations as additional research is conducted on student learning in the coming years.

The learning trajectory displays in the standards charts and hexagon map, as mentioned earlier, must be regarded as highly abridged versions of learning trajectories. They were designed to be as consistent as possible with the CCSS-M standards, in consideration of the importance of the CCSSI for U.S. education. This pragmatic goal served as a design constraint for the displays. The learning trajectory displays and the turnonccmath.net materials provide far more detail on the underlying content and research on mathematics learning for teachers and professional developers than is available in the CCSS-M document itself. The site is extensible, accommodating (1) subsequent development of more detailed and unabridged learning trajectories; (2) revision of the descriptors to further support professional development for implementing the K–8 CCSS-M; (3) links to specific tasks, assessment tools, and curricular referents for improved utility in classrooms; and (4) support of learning progressions-oriented revisions to the standards themselves.

It will be necessary for standards themselves to co-evolve with 21st-century education. The CCSSI anticipated this evolution. The standards will need to be revised in light of the rapidly changing context of education—the changing nature of mathematical sciences, mathematical science-based careers, and their relevance to students' advanced study and career readiness. As research into learning trajectories advances, we will gain new insight into how the meanings that children bring to their learning together with instructional experiences lead to accomplishment of goal-domain knowledge. The learning trajectories and their representations can contribute to the coherence of an instructional core and standards and accountability system that supports children in making the myriad transitions from informal language, strategy, and representation to more formal and flexible uses of concepts and skills in mathematics.

NOTES

1. This chapter's first author was a member of the CCSS-M validation committee; both authors were directly involved in developing drafts of the North Carolina state standards that were developed just prior to the development of the CCSS-M.
2. We note that empirically based learning progressions have not been generated for all of the K–12 mathematical content. The trajectories as displayed in the visual representations described here, whether derived specifically from learning trajectories research or not, are based on our and our colleagues' extensive knowledge of and experience in the relevant mathematical content, curriculum, teaching, and deep grounding in student learning research. They should be considered to be "conjectured." In this chapter, we use the term *learning trajectories* (or *progressions*) to refer to the sequenc-

es of standards that are configured as learning progressions based on our best judgment and that of various colleagues. Our aim in these displays is to lay in a structure organized in a sequence of constructs similar to those be found in empirically supported learning trajectories, and which will promote and accommodate a coherent set of empirically based learning progressions as the ultimate endpoint for the Common Core State Standards for Mathematics.

3. Licensing for distribution was obtained from CCSSO.

REFERENCES

Barrett, J. E., & Clements, D. H. (2003). Quantifying path length: Fourth-grade children's developing abstractions for linear measurement. *Cognition and Instruction, 21*(4), 475–520.

Battista, M. T. (1999). Fifth graders' enumeration of cubes in 3d arrays: Conceptual progress in an inquiry-based classroom. *Journal for Research in Mathematics Education, 30*(4), 417–448.

Battista, M. T. (2007). The development of geometric and spatial thinking. In F. K. Lester (Ed.), *Second handbook of research on mathematics teaching and learning* (pp. 843–908). Charlotte, NC: Information Age Publishing, Inc.

Battista, M. T., Clements, D. H., Arnoff, J., Battista, K., & Borrow, C. V. (1998). Students' spatial structuring of 2d arrays of squares. *Journal for Research in Mathematics Education, 29*(5), 503–532.

Blue Ribbon Commission on Testing and Accountability. (2008). *Report from the blue ribbon commission on testing and accountability to the North Carolina state board of education.* Retrieved from http://www.ncpublicschools.org/docs/acre/history /accountabilityfinalreport.pdf

Brown, A. L. (1992). Design experiments: Theoretical and methodological challenges in creating complex interventions in classroom settings. *Journal of the Learning Sciences, 2*(2), 141–178.

Carpenter, T. P., Franke, M. L., Jacobs, V., Fennema, E., & Empson, S. B. (1997). A longitudinal study of intervention and understanding in children's multidigit addition and subtraction. *Journal for Research in Mathematics Education, 29*(1), 3–30.

Clements, D. H., & Sarama, J. (1992). Early childhood mathematics learning. In D. A. Grouws (Ed.), *Handbook of research on mathematics teaching and learning* (pp. 461–556). New York, NY: Macmillan.

Clements, D. H., & Sarama, J. (2004). Learning trajectories in mathematics education. *Mathematical Thinking and Learning, 6*(2), 81–89.

Clements, D. H., & Sarama, J. (2009). *Learning and teaching early math: The learning trajectories approach.* New York, NY: Routledge.

Clements, D. H., Wilson, D. C., & Sarama, J. (2004). Young children's composition of geometric figures: A learning trajectory. *Mathematical Thinking and Learning, 6*(2), 163–184.

Cobb, P., Confrey, J., diSessa, A. A., Lehrer, R., & Schauble, L. (2003). Design experiments in educational research. *Educational Researcher, 32*(1), 9–13.

Confrey, J. (2006). The evolution of design studies as methodology. In K. R. Sawyer (Ed.), *The Cambridge handbook of the learning sciences* (pp. 135–152). New York, NY: Cambridge University Press.

Confrey, J. (2007, February). *Tracing the evolution of mathematics content standards in the United States: Looking back and projecting forward.* Paper presented at the K–12 Mathematics: What Should Students Learn and When Should They Learn It? conference of the Center for the Study of Mathematics Curriculum, Arlington, VA.

Confrey, J. (2012). Better measurement of higher-cognitive processes through learning trajectories and diagnostic assessments in mathematics: The challenge in adolescence. In V. Reyna, M. Dougherty, S. B. Chapman, and J. Confrey (Eds.), *The adolescent brain: learning, reasoning, and decision making* (pp. 155–182). Washington, DC: American Psychological Association.

Confrey, J., & Maloney, A. P. (2012). Next generation digital classroom assessment based on learning trajectories in mathematics. In C. Dede & J. Richards (Eds.), *Steps toward a digital teaching platform* (pp. 134–152). New York, NY: Teachers College Press.

Confrey, J., Maloney, A. P., Nguyen, K. H., Mojica, G., & Myers, M. (2009). Equipartitioning/splitting as a foundation of rational reasoning using learning trajectories. In M. Tzekaki, M. Kaldrimidou, & H. Sakonidis (Eds.), *Proceedings of the 33rd conference of the international group for the psychology of mathematics education* (Vol. 2, pp. 345–352). Thessaloniki, Greece: International Group for the Psychology of Mathematics Education.

Confrey, J., Nguyen, K. H., Lee, K., Panorkou, N., Corley, A. K., & Maloney, A. P. (2011). *Turn-on common core math: Learning trajectories for the common core state standards for mathematics.* Retrieved from www.turnonccmath.net

Confrey, J., & Scarano, G. H. (1995. October). *Splitting reexamined: Results from a three-year longitudinal study of children in grades three to five.* Paper presented at the annual meeting of the North American Chapter of the International Group for the Psychology of Mathematics Education, Columbus, OH.

Council of Chief State School Officers. (2012). In the States. Retrieved from http://www.corestandards.org/in-the-states

EdWeek Research Center. (2011, Sept 19). No child left behind. *Education Week.* http://www.edweek.org/ew/issues/no-child-left-behind/

Fuson, K. C. (1988). *Children's counting and concepts of number.* New York, NY: Springer-Verlag.

Fuson, K. C. (1992). Research on whole number addition and subtraction. In D. A. Grouws (Ed.), *Handbook of research on mathematics teaching and learning* (pp. 243–275). New York, NY: MacMillan.

Fuson, K. C., Wearne, D., Hiebert, J. C., Murray, H. G., Human, P. G., Oliver, A. I., ... Fennema, E. (1997). Children's conceptual structures for multidigit numbers and methods of multidigit addition and subtraction. *Journal for Research in Mathematics Education, 28*(2), 130–162.

Gardiner, D. P. (1983). *A nation at risk: The imperative for educational reform.* Washington, DC: United States Government Printing Office.

Improving America's Schools Act, P.L. 103-382, 108 Stat. 3518, 1994.

Kelly, A. E. (2008). Reflections on the national mathematics advisory panel final report. *Educational Researcher, 37*(9), 561–564.

Lehrer, R., & Schauble, L. (2006). Cultivating model-based reasoning in science education. In R. K. Sawyer (Ed.), *Cambridge handbook of the learning sciences* (pp. 371–388). Cambridge, MA: Cambridge University Press.

Ma, L. (1999). *Knowing and teaching elementary mathematics: Teachers' understanding of fundamental mathematics in China and the United States.* Mahwah, NJ: Lawrence Erlbaum Associates.

Massell, D. (2008, January). *The current status and role of standards based reform in the states.* Paper presented at the National Research Council Workshop Series on Assessing the Role of K–12 Academic Standards in States, Washington, DC.

National Council of Teachers of Mathematics. (1989). *Curriculum and evaluation: Standards for school mathematics.* Reston, VA: National Council of Teachers of Mathematics.

National Council of Teachers of Mathematics. (2000). *Principles and standards for school mathematics.* Reston, VA: The National Council of Teachers of Mathematics.

National Governors Association Center for Best Practices & Council of Chief State School Officers. (2010). *Common core state standards for mathematics.* Washington, DC: Authors.

National Mathematics Advisory Panel. (2008). *Foundations for success: The final report of the national mathematics advisory panel.* Washington, DC: U.S. Department of Education.

National Research Council. (2004). *On evaluating curricular effectiveness: Judging the quality of K–12 mathematics evaluations.* Washington, DC: The National Academies Press.

National Research Council. (2008). *Common standards for K–12 education?: Considering the evidence: Summary of a workshop series.* Washington, DC: The National Academies Press.

O'Day, J. A., & Smith, M. S. (1993). Systemic reform and educational opportunity. In S. H. Fuhrman (Ed.), *Designing coherent educational policy* (pp. 250–312). San Francisco, CA: Jossey-Bass Publishers.

Porter, A. C., Polikoff, M. S., & Smithson, J. (2009). Is there a de facto national intended curriculum?: Evidence from state content standards. *Educational Evaluation and Policy Analysis, 31*(3), 238–268.

Scarano, G., & Confrey, J. (1996, April). *Results from a three-year longitudinal teaching experiment designed to investigate splitting, ratio and proportion.* Paper presented at the annual meeting of the American Educational Research Association, New York, NY.

Seeley, C. (2008, September). *Calling for a 2020 vision for transforming high school mathematics.* Paper presented at the Future of High School Mathematics Conference, Washington, DC.

Simon, M. A. (1995). Reconstructing mathematics pedagogy from a constructivist perspective. *Journal for Research in Mathematics Education, 26*(2), 114–145.

Standard, *n., adj.* (2007). OED Online. Retrieved from http://www.oed.com/view/Entry/188962?rskey=BtitaY&result=1&isAdvanced=false

Star, S. L. (2010). This is not a boundary object: Reflections on the origin of a concept. *Science Technology & Human Values, 35*(5), 601–617.

Star, S. L., & Griesemer, J. R. (1989). Institutional ecology, 'translation' and boundary objects: Amateurs and professionals in Berkeley's museum of vertebrate zoology, 1907–39. *Social Studies of Science, 19*(3), 387–420.

Steffe, L. P., & Cobb, P. (1988). *Construction of arithmetical meanings and strategies.* New York, NY: Springer-Verlag.

Stevenson, H. W., & Stigler, J. W. (1994). *Learning gap: Why our schools are failing and what we can learn from Japanese and Chinese education.* New York, NY: Simon & Schuster.

Thompson, I. (1999). Written methods of calculation. In I. Thompson (Ed.), *Issues in teaching numeracy in primary schools* (pp. 160–183). Buckingham, UK: Open University Press.

CHAPTER 6

EXPLORING THE RELATIONSHIP BETWEEN LEARNING TRAJECTORIES AND CURRICULUM

A Content Analysis of Rational Number Reasoning in Connected Mathematics and UCSMP Transition Mathematics

Kenny Huy Nguyen and Jere Confrey

ABSTRACT

This chapter considers the relationship between learning trajectories and curriculum through an analysis of the rational number reasoning content of two middle school mathematics curricula (*Connected Mathematics* and *UCSMP Transition Mathematics*) using a learning trajectories perspective. The content analysis revealed that the two curricula approached rational number reasoning differently. *Connected Mathematics* tended toward a "many-to-one" ap-

Learning Over Time: Learning Trajectories in Mathematics Education, pages 161–185.

proach where rational number was developed as ratio, while *Transition Mathematics* tended toward a many-to-one approach where rational number was developed as a point on the number line. Neither curriculum included a full learning trajectories development of rational number as defined by Confrey's (2008) construct map of rational number understanding. Based on the content analysis, an argument is made that research-based learning trajectories are a useful mediational tool that can reveal gaps within a curriculum. Practically, learning trajectories can thus assist teachers to modify their written curriculum to include the missing content, address common misconceptions, and bridge to new content consistent with research in the field of mathematics education.

The term *learning trajectories* has been used in the mathematics education (Clements & Sarama, 2004; Confrey, 2006; Simon, 1995), science education (Corcoran, Mosher, & Rogat, 2009; National Research Council, 2006), and learning sciences (Lehrer & Schauble, 2006) literature to refer to a progression of related concepts that students move through to build knowledge in a domain from informal to more formal concepts of understanding, in a more or less predictable manner over time. They are also referred to in the literature as *conceptual corridors* (Cobb, Confrey, diSessa, Lehrer, & Schauble, 2003), *developmental corridors* (Brown & Campione, 1996), and *learning progressions* (Catley, Lehrer, & Reiser, 2004). While learning trajectories have been used to theoretically inform the work in mathematics education for nearly two decades (Clements & Sarama, 2004; Confrey, 2008; Simon, 1995), only recently has the field begun to empirically validate them and construct assessments to determine where students' proficiencies might lie in relation to a trajectory (Clements, 2010; see also Chapter 2). Our research team at North Carolina State University (DELTA research group) is constructing learning trajectories and assessments for six rational number topics (equipartitioning; division and multiplication; length, area, and volume; fraction; ratio, proportion, and rate; and decimals and percents). Members of our research team have also explored how learning trajectories might help to inform the work of preservice and in-service teachers and, by extension, the progress of students (Mojica, 2010; Wilson, 2009; Wilson, Mojica, & Confrey, 2013). However, a central and unanswered question in the field's examination of learning trajectories is that of the relationship between learning trajectories and curriculum. That is, are learning trajectories different from curriculum and, if so, what are the salient characteristics of these differences? Because both learning trajectories and curriculum seem to posit well-sequenced tasks and experiences that students should encounter to learn the main concepts in a domain, skeptics may question if there are any practical differences between them.

This chapter responds to this broad question by arguing that learning trajectories are indeed different from curriculum: While learning trajectories offer a flexible theory with which to adjust instruction, written curriculum is essentially static.[1] Hence, we see curriculum as a specific set of tasks that lead to learned ideas. In the first part of this chapter, we set up the importance of examining RNR and theorize curriculum as the way in which landmark ideas in a learning trajectory are presented. How this is done matters, and we posit that it should be done consistently with extant research in the field. In the second part of this chapter, we consider the approach of two curricula—*Connected Mathematics* (Lappan, Fey, Fitzgerald, Friel, & Phillips, 1998) and *UCSMP Transition Mathematics* (Usiskin et al., 1998)—to rational number reasoning (RNR) from a learning trajectories perspective. To do this, we describe a content analysis of the curricula that compared their sequencing of RNR topics with Confrey's (2008) construct map of RNR. Neither of these curricula was designed from a rational number learning trajectories perspective, so it would be unreasonable to expect them to align completely with a learning trajectory for RNR or to use this as the basis of evaluation. Instead, the purpose of the content analysis is to highlight how these two curricula's sequencing of rational number topics differ from each other and how they align with and diverge from Confrey's construct map. These differences will then be used to posit how curricula might be improved if teachers viewed them through the lens of a learning trajectory. Learning trajectories, teacher practices, and curriculum are closely interlinked; a theory encompassing all three might help bridge the gap between research and practice, with learning trajectories helping to bring the research base into instructional practice. We conclude with the implications of these ideas for practice and research.

TIMELINESS AND RATIONALE FOR RATIONAL NUMBER

Recent national reports and articles have commented on the poor algebra achievement of students in the United States. Loveless (2008) and the National Mathematics Advisory Panel (2008) suggest that the solution to this problem involves student remediation in basic skills, with a heavy focus on the computational manipulation of fractions. The National Mathematics Advisory Panel (2008), referring to the work of Wu (2007), further suggests focusing on understanding fractions solely as a point on the number line. While we recognize that understanding fraction as a one-dimensional quantity on a number line is important, and that the number line is an elegant representation of the real numbers, we also acknowledge previous research in mathematics education that concludes that understanding fraction as ratio and fraction as operator are equally important ideas (Behr, Harel, Post, & Lesh, 1992; Confrey, Maloney, Nguyen, Wilson, & Mojica, 2008; Kieren, 1992). Confrey (2008) further argues that collapsing a two-dimensional construct like ratio onto the one-dimensional number line results in the loss of rich mathematical meaning. For example, in ratio reasoning, equivalence means an invariance in the relationship of two numbers even as the quantities change. In the case of a familiar early problem in ratio reasoning, if a lemonade recipe calls for one cup of water to 3 lemons, then this recipe would be equivalent to lemonade made with 4 cups of water to 12 lemons (Confrey & Smith, 1995). Although the two recipes contain different amounts of water and lemons, there is something inherently the same about them—the ratio of water to lemons. Reporting on the results of a three-year teaching experiment, Confrey and Scarano (1995) described how third graders understood this notion of "the same," sometimes describing it as the "lemony-ness" of the drink. In the above example, the two ratios of water to lemons would all collapse to 1/3 on the number line, reducing a two-dimensional construct to one dimension.

Perhaps the most difficult challenge for students is that they are expected to fluently negotiate three different underlying constructs that share the fraction symbol as a common representation: fraction as ratio, fraction as operator, and fraction as number (Confrey, 2008). For example, students need to understand that in some circumstances it is perfectly acceptable to add numbers in fractional notation by adding the numerators to each other and the denominators to each other, as when fractional notation is used to represent two ratio units. In the lemonade example above, the fraction 1/3 used to represent 1 lemon to 3 cups of water can be added to itself to get the equivalent ratio 2/6, or 2 lemons to 6 cups of water. However, the context of the notation is critical: This type of addition is not appropriate when the fractions are real numbers.

The National Mathematics Advisory Council (2008) recommended, "A major goal for K–8 mathematics education should be proficiency with fractions (including decimals, percents, and negative fractions), for such proficiency is foundational for algebra" (p. xvii). However, one of its own panel members, the chair of the task group on instructional practices, acknowledged in a presentation at the National Science Foundation that the panel did not find any empirical research to support the claim that success with fractions was linked to success in algebra (Gersten, 2008). This is not to say that fractions are unimportant in algebra, but that they are only one aspect of the entire field of RNR. Previous research in mathematics education, including work by the Rational Number Project (RNP), has instead shown a link between success in algebra and success in rational number, broadly defined (Hiebert & Behr, 1988; Romberg, 1995)—that is, the importance of having a broad understanding of fraction not only as number but also as ratio and operator, along with fluency in multiplication and division, area and volume, decimals and percent, and similarity and scaling. Hence, as Lamon (2007) noted, although young educational researchers are not taking up this important research area because of the necessary long-term research commitments, good research on RNR, taking advantage of new research methods such as the design experiment (Brown, 1992), is needed now more than ever.

Unfortunately, there is a paucity of research that has specifically examined the impact of curriculum on students' RNR abilities; most of this research has come out of the RNP (Behr, Cramer, Harel, Lesh, & Post, 2006). The only major content analysis we found that examined RNR ability as a dependent variable was conducted by the RNP researchers (Cramer, Post, & delMas, 2002), who investigated the impact of their rational number curriculum (taught in 30 weeks), in relation to that of traditional curriculum, on the achievement of 1,600 students on concepts, equivalence, order, and operations with rational numbers as measured by a posttest. Using a MANOVA design, they found an overall main effect in favor of their curriculum and all submeasures, except for addition and subtraction of rational numbers. Follow-up interviews of students revealed that RNP students approached rational number tasks with a conceptually more robust mental image of fractions, whereas traditionally instructed students relied on rote procedures. Despite the study's strong triangulation design and its meeting the National Research Council's (2004) criteria for being *at least minimally methodologically adequate*, some reservations about the study must be noted. First, although the study was randomized at the classroom level, all analyses were conducted at the student level. Second, the sample diversity must be questioned because the classrooms were randomly chosen from a suburban school district south of Minneapolis in which only 1% of students qualified for free and reduced lunch. Third, the study relied only on posttest data as a determination of student achievement. Although the authors justified this decision by arguing that a pretest would have interfered with treatment, we believe that without adequate baseline data as a comparison, it is difficult to interpret the posttest data. These reservations aside, studies that examine curricular effects on RNR ability and curricula written and designed around empirical research are needed in mathematics education.

Another concern is that empirical research on RNR does not seem to have taken root in practice. This is unfortunate because, unlike the National Mathematics Advisory Panel's (2008) contention, the field *does* know a lot about how students reason about and learn rational number (Behr et al., 1992; Lamon, 2007). However, due to a variety of reasons, this research has not permeated into state standards and extant curricula. For example, Moss and Case (1999) argued on the basis of their research that percentages should be sequentially taught before fractions and decimals because of children's early exposure to percentages through social contexts. Confrey (2008) and others (Squire & Bryant, 2002) have argued that division should be sequenced earlier in the elementary school curriculum to take advantage of young children's experiences with sharing. Finally, researchers have argued that the sequencing of arithmetic operations with fractions should be different from that of whole-number operations, because addition of fractions with unlike denominators involves multiplication of fractions (Confrey, 2008; Ni & Zhou, 2005). The above research in RNR makes appeals to children's early experiences, but these innovations have barely made a dent in existing standards and curriculum. In the Common Core State Standards for Mathematics (CCSS-M) (National Governors Association Center for Best Practices & Council for Chief State School Officers, 2010), for example, multiplication and division are still held until the later elementary grades; fraction addition is presented before fraction multiplication; and important mathematical topics like percents, ratios, and data and probability are held until middle school. Therefore, the elephant in the room is this: Why have research results about children's understanding of rational number failed to permeate the curriculum, and what can we do about it? We posit that learning trajectories, above and beyond their uses described in the introduction, could help by creating a more flexible and longitudinal map of student learning that supports teachers in modifying instruction and curricula to align with a robust, empirically supported model of student learning.

FRAMEWORK AND METHODOLOGY

To answer the broader question of the relationship between curriculum and learning trajectories and address the question of how to ensure that empirical research in mathematics education, and especially rational number, is reflected in curriculum, we conducted a curriculum study that sought to examine to what extent two curricula's (*Connected Mathematics* and *UCSMP Transition Mathematics*) sequencing of rational number topics align with Confrey's (2008) rational number construct map. We argue from this analysis that *curriculum matters* and, moreover, that learning trajectories can mediate curriculum, giving teachers guidance on improving the available written curriculum. This section will briefly discuss the framework of learning trajectories, Confrey's rational number construct map, and how the content analysis was conducted.

Learning Trajectories and Curriculum

Our research team has defined a learning trajectory as follows:

> A researcher-conjectured, empirically-supported description of the ordered network of experiences a student encounters through instruction (i.e., activities, tasks, tools, forms of interaction and methods of evaluation), in order to move from informal ideas, through successive refinements of representation, articulation, and reflection, towards increasingly complex concepts over time. (Confrey, Maloney, Nguyen, Mojica, & Myers, 2009, p. 3)

There are four salient components to this definition. First, learning trajectories are researcher-conjectured from empirical evidence in the literature. This suggests that they are conjectured likely passageways that are based on a synthesis—not simply a review—of the literature. Second, it recognizes the role of instruction that includes activities, tasks, tools, forms of interaction, and methods of evaluation. Third, it recognizes that students progress form informal to formal knowledge and that this progression is measurable. Fourth, the definition suggests an iterative process in the construction of learning trajectories, whereby they are continually improved and refined through empirical research.

A learning trajectory can thus be thought of as a progression of related concepts in a domain, examined as a strand across grade levels and carefully sequenced, predicated on certain developmental landmarks, which can help focus instruction on foundational concepts and refine students toward proficiency over time. Students engage in this sequence of related concepts, which become increasingly complex, through engaging in carefully designed tasks with teachers and peers. Learning trajectory topics exhibit predictable patterns of student behaviors, language, misconceptions, representations, justifications, and emergence of mathematical properties. Our research group would further claim that learning trajectories involve significant accommodation or reconceptualization of earlier content through processes of reflective abstraction, generalization, and proof (Simon, 1995).

Confrey (2006) uses the diagram presented in Figure 6.1 to metaphorically describe a learning trajectory as the flow of a stream. In this metaphor, students begin with prior knowledge and travel down the trajectory to learned ideas. The borders of the stream constrain the trajectory to a domain and students progress through landmarks and attempt to circumnavigate obstacles along the trajectory. As described above, landmarks are foundational concepts that are developed and refined over time. These landmarks and obstacles are informed from a synthesis of the research literature in the relevant domain.

In Figure 6.1, a class's progress through the learning trajectory is represented by the black dotted line.[2] Building from this metaphor, we claim that the *flow* down the stream, and especially into a landmark or reckoning with an obstacle, can be represented by the curriculum. Hence, two classrooms approaching a landmark idea through different curricula may experience that landmark differently, and this will affect their progress through the remainder of the trajectory.[3] The content coverage of a curriculum also determines which landmarks students encounter and which ones are skipped. Therefore, we offer a slight modification to Confrey's diagram (Figure 6.2).

In Figure 6.2, two classrooms traverse the same learning trajectory with two different curricula. The progression of Classroom 1 is depicted by the

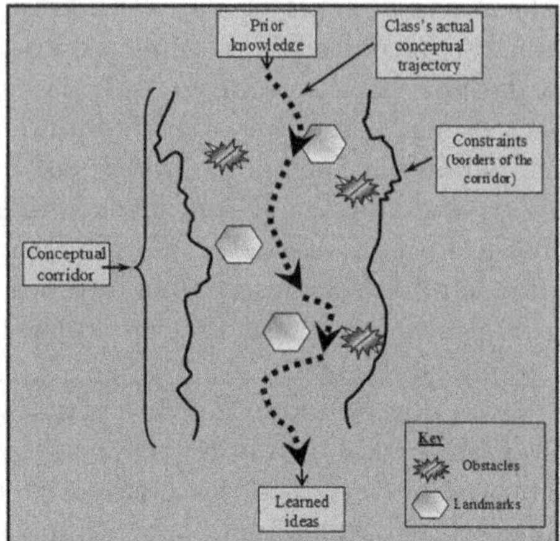

FIGURE 6.1. A conception of a learning trajectory within a conceptual corridor (from Confrey, 2006). *Note:* Please refer to http://gismo.fi.ncsu.edu/LTVolumechapter6 for the color figure.

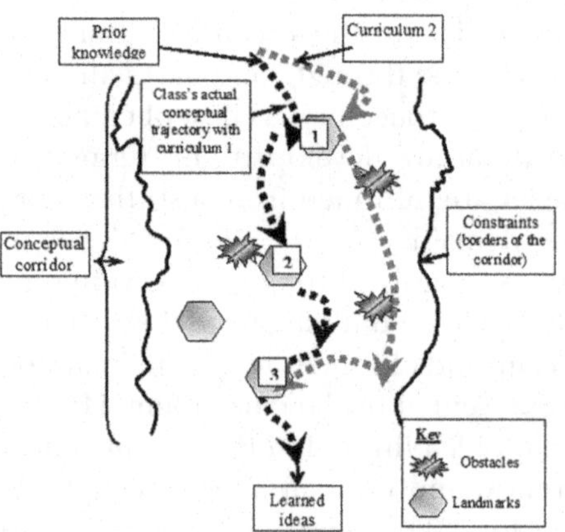

FIGURE 6.2. Learning trajectory and conceptual corridor (Confrey, 2006), amended to include curriculum. *Note:* Please refer to http://gismo.fi.ncsu.edu/LTVolumechapter6 for the color figure.

dotted black line; that of Classroom 2 is depicted by the dotted red line. They both approach Landmark 1, but from two different perspectives. Classroom 1 then proceeds to Landmark 2 and thence to Landmark 3. However, Classroom 2's curriculum skips Landmark 2 and proceeds directly to Landmark 3, and encounters two obstacles that Classroom 1 avoided. Although this is a theoretical example, we show in the following content analysis that such an example is realistic and that it has instructional consequences.

Description of the Content Analysis

The content analysis consisted of analyzing the Grades 6–8 curriculum of the second edition of *Connected Mathematics* (Lappan et al., 1998) and the second edition of *UCSMP Transition Mathematics* (Usiskin et al., 1998) for their coverage and sequencing of RNR topics. Before describing how this was done, we acknowledge that *Connected Mathematics* is a curriculum for three grade levels whereas *Transition Mathematics* is meant as a one-year course whose main purpose is "to provide a smooth path from arithmetic to algebra, and from the visual world and arithmetic to geometry" (Usiskin et al., 1998, p. iv). Because of the different educational time spans and purposes encompassed by these curricula, they are not directly comparable. However, because the intent of *Transition Mathematics* is to bridge arithmetic and algebra, and because its promotional materials claim that it can be used with a diverse range of students (University of Chicago School Mathematics Project, 2008), including students who are performing below grade level, it is reasonable in our view to expect this curriculum to provide an adequate treatment of RNR before students enter *UCSMP Algebra*. We note that although the K–6 curriculum *Everyday Mathematics* was also developed by UCSMP (Bell, Bell, Bretzlauf, Dillard, & Flanders, 2007), it is not assumed to be a prerequisite curriculum for *Transition Mathematics* (C. Siegel, personal communication, December 9, 2008).

In the content analysis, we compared *Connected Mathematics* and *Transition Mathematics* to Confrey's (2008) construct map of RNR (Figure 6.3).

The RNR construct map in Figure 6.3 represents the major topic areas of RNR, including the six topic areas around which the DELTA project is constructing learning trajectories and diagnostic assessments. It also visually represents the three fractional worlds (fraction as ratio, fraction as operator, and fraction as number) alluded to earlier. The RNR construct map incorporates the equipartitioning construct (see Chapter 3, this volume) as a foundational construct that forms a major basis for all three fraction constructs. DELTA project research has demonstrated that the equipartitioning learning trajectory includes foundations for many-as-one, many-to-one, and operator perspectives on fractions. The right side of the map (starting from fair shares and continuing to measure, many-as-one, composite units, etc.) traces the landmarks in the development of the concept of fraction as

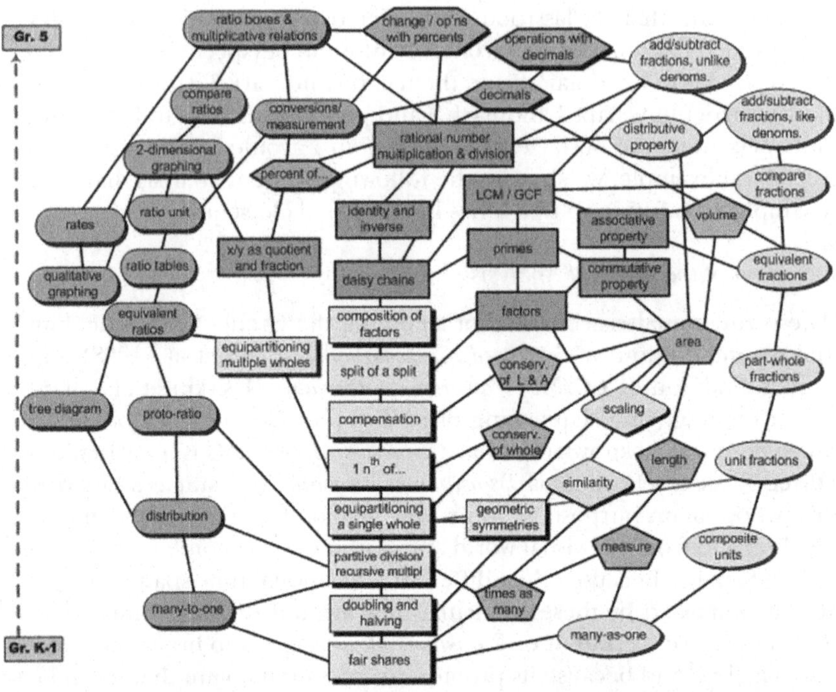

FIGURE 6.3. Rational Number Reasoning construct map of RNR (Confrey, 2008). *Note:* Please refer to http://gismo.fi.ncsu.edu/LTVolumechapter6 for the color-coded map.

a point on the number line; this construct of fraction follows from a many-as-one view of equipartitioning where the answer to the question, "How many coins does each person get if six people share 24 coins?" is four coins (Confrey, 2008). The left side of the map (starting from fair shares and continuing to many-to-one, distribution, etc.) leads the reader through the development of ratio and proportion; this construct of fraction follows from a many-to-one view of equipartitioning in which the answer to the question, "How many coins does each get if six people share 24 coins?" is four coins per person (Confrey, 2008). Finally, fraction as operator can be seen in the multiplicative topics in the middle of the construct map (starting from fair shares and continuing to partitive division, etc.). Confrey (2008) noted that most curricula develop fraction as operator first, pick up length and area, and then heavily develop fraction as a point on the number line. The development of ratio concepts and the role of fraction as ratio is postponed until middle school. This development is reflected in the CCSS-M (CCSSO, 2010). Confrey proposes that student learning of these critical pre-algebraic RNR constructs would be improved considerably if, instead of typical curricular treatments, all three fractional worlds were instructionally developed in parallel. This would support students in earlier understanding and flexible translating among these three fraction worlds. In part, this content analysis will reveal to what extent *Connected Mathematics* and *Transition Mathematics* are consistent with this model for student learning that has been established in the literature (Confrey, 2008; Confrey & Scarano, 1995; Ni & Zhou, 2005; Squire & Bryant, 2002).

For this chapter, we performed a learning trajectories analysis of *Connected Mathematics* and *Transition Mathematics.* This involved two steps. We first examined both curricula and coded the coverage of RNR topics and noted the sequencing of these topics. We then compared the sequencing and coverage of each curriculum with Confrey's (2008) RNR map, which would reveal the extent to which each curriculum aligned with research in the field of RNR. The analysis would also reveal any gaps in the coverage of topics in each curriculum.

We coded coverage only when there was explicit evidence in the written curriculum. That is, we examined the student materials, teacher materials, exercises, assessments, and worksheets. This was particularly challenging with *Connected Mathematics* because this curriculum is presented as a series of complex tasks called *investigations.* Many times, these investigations approach an important rational number topic but fall just short because the written curriculum did not make an explicit connection between the investigation and the topic. For example, while treating an area model based on tile covering, *Connected Mathematics* develops an area model for the distributive property. But nowhere in the written curriculum is the connection between the tiling-area model and the distributive property made explicit to the student or the teacher. Therefore, we did not code that the curriculum covered the distributive property.

After coding each curriculum, we then created separate color-coded maps for each curriculum, using the RNR construct map (Confrey, 2008) as a template to reflect the coverage of topics in the two curricula. The colors ranged from light to dark to indicate the sequencing of topics. For *Connected Mathematics,* we used three hues to show the sequencing across grade levels. For *Transition Mathematics,* only one hue was used.[4]

FINDINGS

The content coverage of RNR topics and sequencing for *Connected Mathematics* in relation to the RNR construct map is presented in Figure 6.4; the RNR content coverage for *Transition Mathematics* is presented in Figure 6.5. For *Connected Mathematics,* sixth-grade topics are coded in blue, seventh-grade topics are coded in green, and eighth-grade topics in red. An increase in color intensity denotes sequencing of topics for each curriculum.

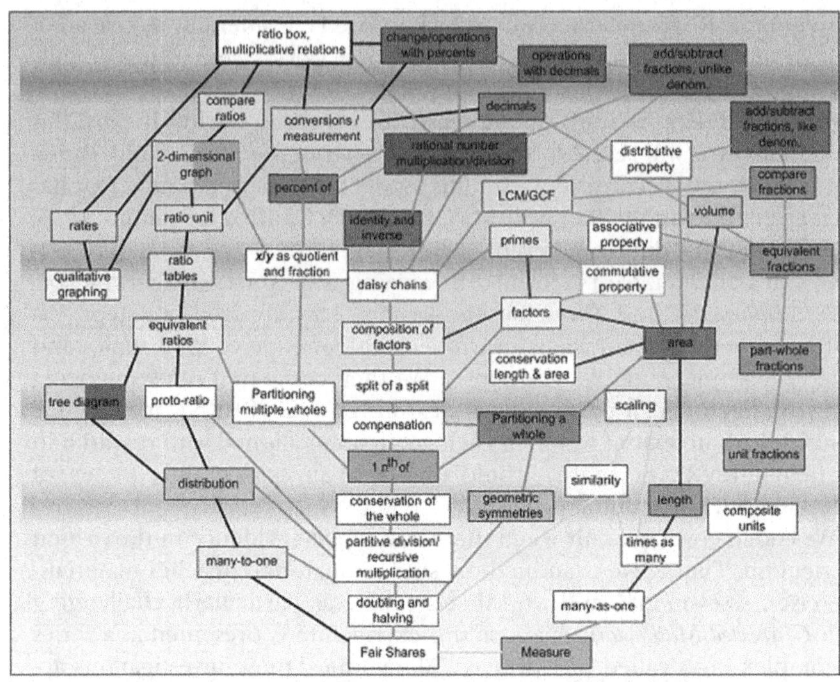

FIGURE 6.4. RNR Constructs and *Connected Mathematics. Note:* Confrey's (2008) construct map coded for sequencing of rational number topics in *Connected Mathematics.* Please refer to http://gismo.fi.ncsu.edu/LTVolumechapter6 for the color-coded map.

Hence, light blue represents topics covered early in the sixth grade curriculum while dark blue represents topics covered late in the sixth grade curriculum. *Transition Mathematics* is coded only in blue, because it is a one-year course. These maps provide a way to visually examine not only each curriculum's coverage of rational number topics but also to compare between curricula. This section briefly reports on the sequencing of each curriculum.

Connected Mathematics

In sixth grade, students begin their coverage of RNR by learning about factors, composition of factors, primes least common multiple (LCM), and greatest common factor (GCF) in the "Prime Time" investigation. This aligns with Confrey's (2008) conjecture that fraction as operator is often encountered first in the curriculum. Students end their investigation of primes by exploring simple tree diagrams as a method of factoring. It is un-

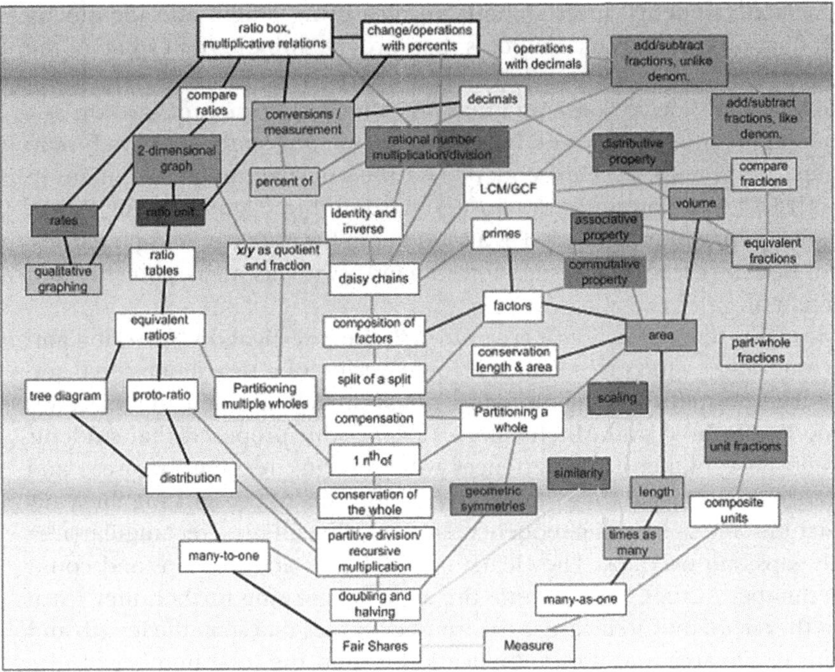

FIGURE 6.5. RNR Constructs and *Transition Mathematics. Note:* Confrey's (2008) construct map coded for sequencing of rational number topics in *Transition Mathematics.* Please refer to http://gismo.fi.ncsu.edu/LTVolumechapter6 for the color-coded map.

fortunate that the Venn diagram model of LCM/GCF was not discussed in this investigation that introduced tree diagrams, Venn diagrams, and LCM/GCF. This is an example of a missed opportunity and what we call a *potentiality* of the curriculum. That is, while an experienced teacher (both mathematically and pedagogically) would be able to make this connection for students, teachers conforming completely to the written curriculum would likely miss this opportunity and move onto the next investigation. Thus, the curriculum misses the opportunity to support both students and less experienced or less knowledgeable teachers by making the connection between Venn diagram models and LCM/GCF.

Two-dimensional graphing and distribution are briefly covered next in an investigation on data. The investigation "Shapes and Designs" then introduces students to geometric symmetry. The next investigation, "Bits and Pieces I," concentrates heavily on fraction as number. In preparing for this sequence, students learn measurement, unit fractions, and fraction as operator by exploring the meaning of $1/n^{th}$ of. These ideas are then used to facilitate students' investigations on measuring length and identifying unit fractions with a ruler. Students then progress up the right spine of the construct map (see Figure 6.3) and learn part–whole fractions, equivalent fractions, and how to compare fractions. After a discussion of fraction as a point on the number line and fraction as an operator, students learn how to equipartition a whole. This sequence is different than the development in the DELTA learning trajectory for equipartitioning (Confrey et al., 2008), where equipartitioning is used as a foundation for understanding fractions. The "Bits and Pieces I" investigation concludes with decimals, percents, and percent of.

Students then learn about area through the investigation "Covering and Surrounding," which treats area as the covering of a two-dimensional surface. However, this investigation does not approach this topic the way it was done by Outhred and Mitchelmore (2000), who proposed that students should have numerous experiences with covering rectangular areas and come to an understanding of the $A = l \times w$ formula by recognizing the array structure. In that learning sequence, students first fill-tile a rectangular area with gaps and overlaps. They learn to avoid gaps and overlaps and count the number of tiles to determine the area. Progressing further, they learn that they need only to measure the number of tiles that span the length and width of the area and use multiplication to find the total number of tiles rather than counting or repeated addition.

Connected Mathematics approaches area by first having students investigate covering *irregular* rectangular areas through estimation, followed by some work with equipartitioning of the areas, which is likely meant to help students understand the continuous nature of the iterable unit. Students then learn the $A = l \times w$ formula not through a sequence of tasks leading to the formation of an array as suggested by Outhred and Mitchelmore (2000) and Clements and Sarama (2009), but by making tables of the relationship among length, width, and area of many rectangular areas and recognizing $A = l \times w$ as an algebraic pattern.

After an investigation on probability, students learn the remainder of the topics in the fraction-as-number branch of the RNR map (Confrey, 2008) and use these ideas to ultimately extend multiplication and division to rational numbers in the investigation "Bits and Pieces II." They first learn to add and subtract fractions, beginning with fractions that have common denominators; perform addition and subtraction with decimals; learn to perform operations with percents; convert among fractions, decimals, and percents; and finally learn multiplication and division of rational numbers.

Investigating and learning about ratios is the major goal of the seventh-grade *Connected Mathematics* curriculum. After an initial investigation of variables and patterns, students investigate similarity and scaling. Ratio is then covered in depth, beginning with understanding the many-to-one relationship and then proceeding to cover topics in the left spine of the RNR map (Confrey 2008) including equivalent ratios, ratio tables, ratio units, comparing ratios, and rates. Students' ratio knowledge is then applied to help them convert quantities within the same measurement system. The seventh-grade curriculum concludes with an investigation on volume.

Very few RNR topics are covered in eighth grade *Connected Mathematics*—most of the eighth-grade curriculum centers on algebra. Students first revisit graphing in order to graph linear equations. Area and volume are then covered in more detail, and students learn area and volume formulas for a variety of geometric objects beyond simple squares, rectangles, and cubes. Identity and inverse are then covered. Finally, tree diagrams are revisited in the context of combinatorics.

Transition Mathematics

Transition Mathematics begins its coverage of RNR topics with measure and decimals, and then intensively covers x/y as quotient and fraction. These curricular moves set up the rest of the curriculum to handle fraction as a point on the number line. Fractions are covered after the equivalence of quotient and fraction are established, and students learn the topics along the part of the right spine of the RNR map (Confrey, 2008), covering part–whole fractions, comparison of fractions, and equivalent fractions. Percent and "percent of" are then covered as well.

The curriculum then proceeds to qualitative graphing and the *times as many* concept, leading into a discussion of length, and then unit conversions, which are handled as a context for measurement. This progression is unlike the one in *Connected Mathematics* that covered conversions only after a careful discussion of ratio. Students then learn area and volume in the more traditional manner of applying the $A = l \text{ x } w$ formula to different problem contexts. Students then move to adding and subtracting fractions, and unit fractions, in a very perfunctory exercise-based manner. Two-dimensional graphing and linear equations are then covered, along with a brief interlude in geometric symmetry. Students then are introduced to rational number multiplication and work with the associative and commutative properties of addition and multiplication.

Rate is introduced with only a single lesson with no previous work on ratio. Rate is defined simply as "a quantity is a rate when its unit contains the word 'per' or 'for each' or some synonym" (Usiskin et al., 1998, p. 493). The curriculum concludes with a discussion of similarity and scaling, the distributive property, rational number division, and ratio unit. Ratio units are used only for "ratio comparisons," which leads to solving ratio and proportion problems with the traditional method of equating the product of the means with the product of the extremes.

Like rate, ratio is not heavily developed and is defined as follows:

> Notice what happens when the units are put into the numerator and denominator. 11 days / 30 days. The units are the same. They cancel in the division, so the answer has no unit. Therefore, this is not an example of a rate. The answer is an example of ratio. Ratios have no units. Because the 11 days is being compared to the 30 days, this use of division is called ratio comparison. (Usiskin et al., 1998, p. 611)

This definition is problematic in light of the research literature in rational number, which will be discussed below.

DISCUSSION

Comparison of Figures 6.4 and 6.5 demonstrates that *Connected Mathematics* and *Transition Mathematics* differ greatly in their approaches to RNR. Although *Connected Mathematics* begins with fraction as a point on the number line, this idea is gradually built up using equipartitioning, 1/*n*th of, and unit fractions. *Connected Mathematics* then rigorously develops the concepts of ratio and rate instead of relying on both a perfunctory definition and simply teaching students to calculate the unit ratio or rate in order to solve problems involving ratio and rates. The constructs mapped for *Connected Mathematics* show a more thorough coverage of the three fractional worlds.

Transition Mathematics takes a very different approach, beginning immediately with decimals and emphasizing x/y as quotient and fraction as number. The number line is also introduced in the first chapter of *Transition Mathematics* to help students locate decimals and fractions on the real line. Perhaps most problematic from the perspective of the RNR map (Confrey, 2008), fraction as ratio and fraction as operator are not considered in the *Transition* curriculum. As discussed in the previous section, *Transition Mathematics* defines rates only perfunctorily and semantically—simply as a quantity with "per" or "for each"—and ratios as dimensionless fractions used for comparison purposes.

We claim that in light of Confrey's work on ratio (Confrey, 2008; Confrey & Scarano, 1995; Confrey, Nguyen, Maloney, Lee, Corley, & Panorkou, 2011), these definitions of rate and ratio and curtailing the treatment of ratio to the unit ratio only fail to account for the complexity of the ratio world. For example, Confrey et al. (2011) note that in general for two quantities *a* and *b* that are in a ratio relationship, curriculum and standards generally only deal with the unit ratio a/b : 1 and treat it as *the* unit ratio hence missing the second unit ratio b/a : 1. The importance of both unit ratio types is that, in actuality, either unit ratio can be used to find the missing value in a ratio problem (even though in practice usually only a/b : 1 is used). Ignoring the second unit is analogous to confining ratio, a two-dimensional construct, to a one-dimensional construct. Although we acknowledge that curriculum and standards writers do this to simplify the process of working with ratios as operators, it hides the two-dimensional nature of ratios. We argue that percentages as operators will subsequently be much easier to establish instructionally if both unit ratios are acknowledged.

The treatments of ratio in *Connected Mathematics* and *Transition Mathematics* comprise an example of the illustration of multiple curricular paths through a conceptual corridor, as diagrammed in Figure 6.2. The comparison of these two curricular treatments provides an example of why curriculum matters. Both curricula endeavor to develop ratio as a way to solve problems involving ratio, proportion, and rate. However, their methods of developing ratio are quite different. *Connected Mathematics'* students develop equivalent ratios, ratio tables, and ratio units, whereas *Transition Mathematics* skips these landmarks and expects students to learn unit ratio as a tool to solve ratio problems using the so-called "means-extremes" property (Usiskin et al., 2008, p. 620). We argue that this sets up potential obstacles for students, such as failure to understand the differences in the use of fractional notation to denote a "real number" (fraction) and a ratio, and not understanding that ratio is a two-dimensional construct. It may be argued that because *Transition Mathematics* is a one-year curriculum, it should not be expected to extensively cover ratio to the extent done in *Connected Mathematics,* a three-year curriculum. However, as Confrey (2008) articulated in her plenary address for the International Congress on Mathematics Education, students are generally not exposed to ratio in elementary school: For example, the previous state curricula for both North Carolina and Missouri (North Carolina Department of Public Instruction, 2003; Missouri Department of Elementary and Secondary Education, 2004) and the CCSS-M (CCSSO, 2010) postpone directly addressing ratio topics until middle school. An examination of the next textbook in the UCSMP series, *UCSMP Algebra* (McConnell, Brown, & Usiskin, 1998), did not reveal evidence of these missing ratio landmarks. Hence, students who use *Transition Mathematics* without a solid ratio background are unlikely to become proficient in ratio reasoning in *Transition Mathematics* or future *UCSMP* textbooks without specific changes in or supplementation of the written curriculum. This concern raises an important issue for learning trajectories. As researchers, how do we help inform teachers about landmarks missing from a student's learning profile? We will argue below that valid diagnostic assessments can provide an answer to this question.

The final step in our analysis was to examine the gaps in each curriculum's coverage of RNR topics, to determine with which topics teachers should supplement these curricula to be consistent with the research literature. It is reasonable to expect that equipartitioning would not be covered in these middle grades curricula; nonetheless, other gaps raise some concerns. In general, besides providing a more complete coverage of the three fractional worlds, *Connected Mathematics* covers more of the rational number construct map than *Transition Mathematics.* It is nevertheless interesting to note the strengths and weaknesses of each curriculum's coverage. For example, *Transition Mathematics* does a complete coverage of the associative, commutative, and distributive properties, providing students with formal statements and applications of each property and expecting them to use these properties to justify the manipulation of algebraic equations. However, there was no evidence of discussion of these properties in the *Connected Mathematics* curriculum. When discussing an area model for multiplication, the curriculum came close to discussing a visual model for the distributive property but, like many topics, the investigation never made this connection clear; therefore, this must be filed as another potentiality of the curriculum.

Surprisingly absent from *Transition Mathematics* is coverage of factors, primes, and LCM/GCF, three major topics that form a basis for proficiency in multiplicative reasoning (Harel & Confrey, 1994). The absence of these topics from this curriculum suggests that if students have subsequent difficulties in algebra, their challenges should not be attributed primarily to poor skills in basic arithmetic and symbolic manipulation, but rather to their lack of opportunity to learn to move flexibly in multiplicative space (that is, to understand the structure of number from a multiplicative perspective—the *biography* of a number as its factors) and among the three fractional worlds.

In this section we have discussed how both examining the sequencing of topics and identifying the gaps in a curriculum's treatment of rational number reveals its stance on rational number. *Connected Mathematics* painted a more nuanced picture of rational number covering fraction as ratio, operator, and number. In contrast, *Transition Mathematics* started with fraction as number and used this as its dominant paradigm; when ratios were later introduced, they were immediately collapsed down to decimal notation for use in a "ratio comparison model."

Content analysis of an investigations-based curriculum is challenging, if many mathematical ideas are not explicit in the written curriculum and if teachers are provided insufficient guidance in the supporting materials to give teachers confidence that these connections would be made explicit to students in the enacted curriculum. It may therefore be advisable to consider learning trajectories as a tool to support teachers in supplementing existing curricula to fill in gaps (such as in *Transition Mathematics*) or in connecting "big ideas" (in an investigations-based curriculum). In the Conclusions section, we discuss this idea and give thoughts for future research.

CONCLUSIONS AND IMPLICATIONS FOR FUTURE RESEARCH

We argue that the main differences between the written curriculum and learning trajectories is that while the written curriculum is essentially static, a learning trajectory is more robust because it is the articulation of a theory that provides teachers information about a wide set of student behaviors and can support numerous ways to sequence tasks and experiences to lead to more sophisticated student knowledge over time. However, the theoretical nature of learning trajectories may make them difficult for teachers to put into practice. We believe that a major potential of learning trajectories in relation to curriculum is to act as a mediator, supporting teachers to understand where gaps occur in their curriculum and where they might diverge from or supplement the curriculum to best help their students. A teacher using a learning trajectories lens should be able to supplement the written curriculum, turning it into a more effective enacted curriculum. This would have the added benefit of ensuring that the research into students' mathematics learning becomes more integrated into instructional practice. Introducing teachers, through professional development efforts, to examination of their curriculum with a learning trajectories lens is a feasible goal. As Penuel, Confrey, Maloney, and Rupp (2013) discuss in relation to the DELTA project, if such a practice could be sustained in a school through a communities of practice approach, it would be a powerful method of enacting change in schools.

We summarize our conclusions and implications for future research as follows:

1. Any curriculum is static, presenting an ordered sequence of tasks that leads students through a domain. Teachers who strictly follow a curriculum have few choices to make—the curriculum typically dictates what topics and tasks students engage with, when, and in what order. Learning trajectories support teachers to make hypotheses about student learning and to modify instruction appropriately to reflect what they understand about their students' past experiences and their current classroom achievement. Although they offer an ordered network of experiences, a learning trajectories theory should guide teachers to several different options after a landmark has been encountered and learned. Assessment is the key to finding out if a landmark has been learned and how to focus instruction to take students to the next set of landmarks.
2. Curriculum can also be thought of as how students encounter an idea in a learning trajectory. How a classroom encounters a landmark (i.e., what curriculum is used) supports or hinders students' learned ideas. However, given the static nature of a curriculum, teachers who strictly follow the curriculum will often proceed to the next topic without reflection on the larger theory. Therefore, teachers should have more guidance on interpreting and supplementing the written curricula to support student understanding.

3. Learning trajectories can provide a mediational tool for teachers to use to view curriculum, supplement it, and, therefore, modify it into a much more effective enacted curriculum. Curriculum could represent the curriculum author's view of a learning trajectory, but it is a static view. Teachers who strictly follow a written curriculum cannot adapt and change by filling in gaps or connecting big ideas, two of the weaknesses evident from the content analysis of *Connected Mathematics* and *Transition Mathematics.* Teachers who use learning trajectories to mediate these curricula may be able to shore up each curriculum's weaknesses in rational number reasoning.

The above claim is large in scope and will require empirical testing. However, it is an important question to consider because individual teachers generally have little say in choosing their curriculum, while being individually responsible for instruction that relies on the curriculum. If it can be shown that teachers who have an understanding of learning trajectories can and do actively modify their curricula in light of the learning trajectory, it would behoove us to promote the theory. In such a circumstance, doing so would not only improve classroom instruction but also ensure that research is actively used to inform instruction. Obviously, "preparing the ground" in a state by modifying state descriptors or having influence on how topics are assessed are other ways to have impact on the system.

In the end, assessment is the unaddressed issue that has quietly lurked in the background of this chapter. Teachers need feedback on what students know, which landmarks they have achieved, how deeply they understand these landmarks, and which landmarks they have missed in their educational careers in order to adapt their instruction to the progress, achievement, and needs of their students. In the end, a learning trajectories-based approach will rely on valid diagnostic measures to "check" that students are indeed learning the landmark ideas. Such assessments may make sequencing less important and perhaps make way for more innovative curricula that emphasize modeling. That is, we believe that part of the problem with our handling of the "misplaced tenth" (Loveless, 2008) is that we track them into remedial courses that teach content by repeating the curriculum that students already failed to learn or simply "teach to the test." We can best serve the needs of struggling students, including urban youth, by attracting them to mathematics through engagement with compelling problems that matter to them in their everyday social experiences (Confrey, Maloney, Ford, & Nguyen, 2006; Lehrer, Kim, & Schauble, 2007). But we must make the case to policymakers and administrators that these curricula "work." One way to convince them of the effectiveness of such methods would be to show that remediation based around an innovative technological or modeling curriculum yields better assessment benefits than traditional remediation courses. An intervention carefully and precisely linked with a diagnostic model as a means to validate that approach could help build effective remediation models.

We conclude by acknowledging the large amount of work that needs to be done to test the claims and hypotheses presented in this paper. One question that has arisen during the work on learning trajectories is how, exactly, they should be presented to teachers. Should it be theoretical in nature, in the form of progress variables (Kennedy & Wilson, 2007), diagnostic assessments, or in the form of a complete curriculum constructed from a learning trajectories-based approach? Wilson's (2009) work has shown that without proper professional development, teachers will use a learning trajectory as simply a "checklist" of behaviors they should be looking for and will not be equipped to provide the scaffolding needed to move students from naïve conceptions to more complicated mathematical reasoning. The DELTA project's current answer to this question is to tie the presentation of learning trajectories into an unpacking of the CCSS-M. We have created a Web site (http://www.turnonccmath.com) where users can explore 18 different trajectories in grades K–8 in the CCSS-M in a "hexagon map" (Confrey et al., 2011). Clicking on any individual trajectory brings up a window that "unpacks" the standards in each trajectory. Beyond unpacking the trajectory, we have also added and unpacked "bridging" standards that are supported and informed by the research literature. We believe that this is a start to presentation of learning trajectories in a way that is both useful to teachers and is aligned to CCSS-M. We believe that this tool and the overall approach will assist teachers to perform learning trajectories-based analyses of their curricula and, therefore, help them to create a more powerful enacted curriculum that is also supported in the CCSS-M. We invite other content experts in mathematics education to do the same to assist teachers in this exciting new age of curricular and standards reform.

NOTES

1. Here, we distinguish between the written curriculum and the enacted curriculum (Tarr, Reys, Reys, Chávez, Shih, & Osterlind, 2008). In this chapter, curriculum will refer to the written curriculum, which, similarly to the National Research Council's (2004) definition, we define to consist of a set of student materials (i.e., the textbook), teacher guides, and accompanying classroom assessments. Any teacher modifications to these materials are not considered to be part of the written curriculum. For example, a teacher who supplements an investigations curriculum lesson with practice problems has modified the written curriculum into her enacted curriculum.

2. Due to publication limitations, all figures appear in black and white. Web addresses for color versions are provided in figure captions.
3. For the purposes of this example, we assume here that the classroom teacher is implementing the written curriculum exactly as intended by the curriculum authors.
4. As noted previously, all figures appear in black and white. Web addresses for color versions are provided in figure captions.

REFERENCES

Behr, M., Cramer, K. A., Harel, G., Lesh, R., & Post, T. (2006). *The rational numbers project* Retrieved from http://cehd.umn.edu/rationalnumberproject/default.html

Behr, M. J., Cramer, K., Harel, G., Lesh, R., & Post, T. (2008). *Rational number project.* University of Minnesota. Retrieved December 2, 2008, from http://www.cehd.umn.edu/ci/rationalnumberproject/

Behr, M., Harel, G., Post, T., & Lesh, R. (1992). Rational number, ratio and proportion. In D. Grouws (Ed.), *Handbook of research on mathematics teaching and learning* (pp. 296–333). New York, NY: Macmillan.

Bell, M., Bell, J., Bretzlauf, J., Dillard, A., & Flanders, J. (2007). *Everyday mathematics.* New York, NY: McGraw-Hill/Glencoe.

Brown, A. L. (1992). Design experiments: Theoretical and methodological challenges in creating complex interventions in classroom settings. *The Journal of the Learning Sciences, 2*(2), 141–178.

Brown, A. L., & Campione, J. (1996). Psychological theory and the design of innovative learning environments: On procedures, principles, and systems. In L. Schauble & R. Glaser (Eds.), *Innovations in learning: New environments for education* (pp. 289–298). Mahwah, NJ: Lawrence Erlbaum Associates.

Catley, K., Lehrer, R., & Reiser, B. (2004). *Tracing a prospective learning progression for developing understanding of evolution.* Washington, DC: National Academy Press.

Clements, D. (2010, May). *Evaluation of a developmental progression for length measurement using the rasch model.* Paper presented at the annual meeting of the American Education Research Association, Denver, CO.

Clements, D. H., & Sarama, J. (2004). Learning trajectories in mathematics education. *Mathematical Thinking and Learning, 6*(2), 81–89.

Clements, D. H., & Sarama, J. (2009). *Learning and teaching early math: The learning trajectories approach.* New York, NY: Routledge.

Cobb, P., Confrey, J., diSessa, A. A., Lehrer, R., & Schauble, L. (2003). Design experiments in educational research. *Educational Researcher, 32*(1), 9–13.

Confrey, J. (2006). The evolution of design studies as methodology. In K. R. Sawyer (Ed.), *The Cambridge handbook of the learning sciences* (pp. 135–152). New York, NY: Cambridge University Press.

Confrey, J. (2008, July). *A synthesis of the research on rational number reasoning: A learning progressions approach to synthesis.* Paper presented at the 11th International Congress of Mathematics Instruction, Monterrey, Mexico.

Confrey, J., Maloney, A. P., Ford, L., & Nguyen, K. H. (2006, December). *Graphs 'n Glyphs as a means to teach animation and graphics to motivate proficiency in mathematics by middle grades urban students.* Paper presented at the ICMI XVII Study, Hanoi, Vietnam.

Confrey, J., Maloney, A. P., Nguyen, K. H., Mojica, G., & Myers, M. (2009). Equipartitioning/splitting as a foundation of rational reasoning using learning trajectories. In M. Tzekaki, M. Kaldrimidou, & H. Sakonidis (Eds.), *Proceedings of the 33rd conference of the International Group for the Psychology of Mathematics Education* (Vol. 2, pp. 1–8). Thessaloniki, Greece: International Group for the Psychology of Mathematics Education.

Confrey, J., Maloney, A. P., Nguyen, K. H., Wilson, P. H., & Mojica, G. (2008, April). *Synthesizing research on rational number reasoning.* Paper presented at the National Council of Teachers of Mathematics Research Presession, Salt Lake City, UT.

Confrey, J., Nguyen, K. H., Maloney, A. P., Lee, K., Corley, A., & Panorkou, N. (2011). Turn on common core math. Retrieved from http://www.turnonccmath.com

Confrey, J., & Scarano, G. H. (1995, October). *Splitting reexamined: Results from a three-year longitudinal study of children in grades three to five.* Paper presented at the Annual Meeting of the North American Chapter of the International Group for the Psychology of Mathematics Education, Columbus, OH.

Confrey, J., & Smith, E. (1995). Splitting, covariation and their role in the development of exponential functions. *Journal for Research in Mathematics Education, 26*(1), 66–86.

Corcoran, T., Mosher, F. A., & Rogat, A. (2009). *Learning progressions in science: An evidence-based approach to reform.* New York, NY: Center on Continuous Instructional Improvement Teachers College.

Cramer, K. A., Post, T. R., & delMas, R. C. (2002). Initial fraction learning by fourth- and fifth-grade students: A comparison of the effects of using commercial curricula with the effects of using the rational number project curriculum. *Journal for Research in Mathematics Education, 33*(2), 111–144.

Gersten, R. (2008). *The gaping holes in the national mathematics advisory panel report.* Paper presented at the NSF Discovery Research Conference, Washington, DC.

Harel, G., & Confrey, J. (Eds.). (1994). *The development of multiplicative reasoning in the learning of mathematics.* Albany, NY: State University of New York Press.

Hiebert, J., & Behr, M. (Eds.). (1988). *Number concepts and operations in the middle grades.* Reston, VA: National Council of Mathematics Teachers.

Kennedy, C. A., & Wilson, M. (2007). Using progress variables to map intellectual development. In R. W. Lissitz (Ed.), *Assessing and modeling cognitive development in schools: Intellectual growth and standard setting.* Maple Grove, MN: JAM Press.

Kieren, T. E. (1992). Rational and fractional numbers as mathematical and personal knowledge: Implications for curriculum and instruction. In G. Leinhardt, R. Putnam, & R. Hattrup (Eds.), *Analysis of arithmetic for mathematics teaching* (pp. 323–372). Hillsdale, NJ: Lawrence Erlbaum Associates.

Lamon, S. J. (2007). Rational numbers and proportional reasoning: Toward a theoretical framework. In F. K. Lester (Ed.), *Second handbook of research on mathematics teaching and learning* (pp. 629–668). Charlotte, NC: Information Age Publishing, Inc.

Lappan, G., Fey, J. T., Fitzgerald, W. M., Friel, S. N., & Phillips, E. D. (1998). *Connected mathematics.* Menlo Park, CA: Dale Seymour.

Lehrer, R., Kim, M., & Schauble, L. (2007). Supporting the development of conceptions of statistics by engaging students in measuring and modeling variability. *International Journal of Computers for Mathematical Learning, 12*(3), 195–216.

Lehrer, R., & Schauble, L. (2006). Cultivating model-based reasoning in science education. In R. K. Sawyer (Ed.), *Cambridge handbook of the learning sciences* (pp. 371–388). Cambridge, UK: Cambridge University Press.

Loveless, T. (2008). *The misplaced math student: Lost in eighth-grade algebra.* Washington, DC: The Brookings Institute.

McConnell, J. W., Brown, S., & Usiskin, Z. (1998). *UCSMP algebra.* Glenview, IL: Scott Foresman/Addison-Wesley.

Missouri Department of Elementary and Secondary Education. (2004). Mathematics grade-level expectations Retrieved from http://dese.mo.gov/divimprove/curriculum/GLE/Math GLE FINAL-ALL SECTIONS.DOC

Mojica, G. (2010). *Preparing pre-service elementary teachers to teach mathematics with learning trajectories.* Unpublished doctoral dissertation, North Carolina State University, Raleigh, NC.

Moss, J., & Case, R. (1999). Developing children's understanding of the rational numbers: A new model and an experimental curriculum. *Journal for Research in Mathematics Education, 30*(2), 122–147.

National Governors Association Center for Best Practices & Council of Chief State School Officers. (2010). *Common core state standards for mathematics.* Washington, DC: Authors.

National Mathematics Advisory Panel. (2008). *Foundations for success: The final report of the national mathematics advisory panel.* Washington, DC: U.S. Department of Education.

National Research Council. (2004). *On evaluating curricular effectiveness: Judging the quality of k-12 mathematics evaluations.* Washington, DC: The National Academies Press.

National Research Council. (2006). *Systems for state science assessment.* Washington, DC: National Academies Press.

Ni, Y., & Zhou, Y.-D. (2005). Teaching and learning fractions and rational numbers: The origins and implications of whole number bias. *Educational Psychologist, 40*(1), 27–52.

North Carolina Department of Public Instruction. (2003). North Carolina standard course of study. Retrieved from http://www.dpi.state.nc.us/curriculum/mathematics/scos/2003/k-8/index

Outhred, L., & Mitchelmore, M. (2000). Young children's intuitive understanding of rectangular area measurement. *Journal for Research in Mathematics Education, 31*(2), 144–167.

Penuel, W. R., Confrey, J., Maloney, A. P., & Rupp, A. A. (2013). Design decisions in developing assessments of learning trajectories: A case study. *Journal of the Learning Sciences,* DOI: 10.1080/10508406.2013.866118

Romberg, T. A. (Ed.). (1995). *Reform in school mathematics and authentic assessment.* Albany, NY: SUNY Press.

Simon, M. A. (1995). Reconstructing mathematics pedagogy from a constructivist perspective. *Journal for Research in Mathematics Education, 26*(2), 114–145.

Squire, S., & Bryant, P. (2002). The influence of sharing on children's initial concept of division. *Journal of Experimental Child Psychology, 81*(1), 1–43.

Tarr, J. E., Reys, R. E., Reys, B. J., Chávez, O., Shih, J., & Osterlind, S. J. (2008). The impact of middle-grades mathematics curricula and the classroom learning environment on student achievement. *Journal for Research in Mathematics Education, 28*(3), 195–213.

University of Chicago School Mathematics Project. (2008). UCSMP home page. Retrieved from http://ucsmp.uchicago.edu/

Usiskin, Z., Feldman, C. H., Davis, S., Mallo, S., Sanders, G., Witonsky, D., ... Viktora, S. S. (1998). *UCSMP transition mathematics.* Glenview, IL: Scott, Foresman and Company.

Wilson, P. H. (2009). *Understanding the effects of a learning trajectory for equipartitioning in classrooms: A mixed methods investigation.* Unpublished doctoral dissertation. North Carolina State University, Raleigh, NC.

Wilson, P. H., Mojica, G., & Confrey, J. (2013). Learning trajectories in teacher education: Supporting teachers' understandings of student thinking. *The Journal of Mathematical Behavior, 32*(1), 103–121.

Wu, H. (2007). Fractions, decimals, and rational numbers Retrieved from http://math.berkeley.edu/~wu/

CHAPTER 7

ARTICULATED LEARNING TRAJECTORIES

A Framework for Mathematical Content Analyses

Travis A. Olson

ABSTRACT

The research reported in this chapter is based on a mathematical content analysis using an *articulated learning trajectory* (ALT) framework, used here to identify the textual development of algebraic thinking concepts through instances of content involving patterning constructs in four middle-school textbook series—*Connected Mathematics 2* (Lappan, Fey, Fitzgerald, Friel, & Phillips, 2009), *Mathematics: Applications and Concepts* (Bailey et al., 2006), *MathThematics* (Billstein & Williamson, 2008), and *Saxon Math* (Hake, 2007). The ALT framework is discussed and delineated as it pertains to this study, and key findings are presented. Differences were found in the number of instances involving patterning constructs and contexts, in the textual placement and emphasis of particular patterning constructs and contexts, and in the development of algebraic thinking concepts. Working visual models for ALTs are presented, and implications are discussed for studies and profession-

Learning Over Time: Learning Trajectories in Mathematics Education, pages 187–226.

al development regarding the implemented curriculum, learning trajectory-based studies of students' knowledge development, alignment studies with respect to the learning trajectories within the *Common Core State Standards for Mathematics,* and the use of the ALT framework in mathematics teacher education programs.

Mathematics textbooks play an important role in teaching and learning mathematics. Teachers rely on curriculum materials (textbooks, teacher editions, workbooks, supplemental materials, etc.) as the foremost resource in determining the mathematics they teach (Chávez, 2003; Clements, 2002; Grouws & Cebulla, 2000; Grouws, Smith, & Sztajn, 2004; McNaught, Tarr, & Grouws, 2008; Mullis et al., 2000; Tarr, Chávez, Reys, & Reys, 2006; Tyson & Woodward, 1989; Woodward & Elliot, 1990). There is also strong indication that teachers supplement textbooks considerably; there have been recent efforts to understand the substantial use of supplemental materials by teachers (Drake, 2010; Huntley & Chval, 2010; McNaught et al., 2008). With or without supplementation, textbooks are vitally important in determining the scope and sequence of the mathematics content within the classroom.

In a recent summary report of a conference examining curriculum design, development, and implementation in the current era of the Common Core State Standards (National Governors Association Center for Best Practices and Council for Chief State School Officers, 2010), Confrey and Krupa (2010) note, "The expert group of designers and publishers of mathematics curriculum materials...along with front-line implementers from states and large school districts...and policy experts from major funding organizations in Washington DC.... began with a shared fundamental assumption: Curricula matter!" (p. 3). The delineation of mathematical content in textbooks influences the way in which teachers select, sequence, and present mathematical "tasks or problems that students are asked to solve," some of the key elements typically associated with a "learning ecology" (Cobb, Confrey, diSessa, Lehrer, & Schauble, 2003, p. 9). However, very little is known regarding key differences among curriculum features, the development of content within curriculum materials, and distinctions that can be made among curricular programs,[1] in particular about the scope and sequence of mathematical content within and among textbook materials currently used by teachers (Remillard, 2009). In the absence of understanding these key differences, there are many questions concerning the implications of the use of curricular materials for student learning. For example, understanding whether there are "shortcomings in the way the content in the textbook is presented" (McNaught et al., 2008, p. 13) could provide deeper understanding of teachers' supplementation of textbooks.

In this chapter, I first discuss three theoretical considerations: (1) defining a construct to frame mathematical content analyses based in the literature of learning trajectories, namely, an *articulated learning trajectory* (ALT); (2) positioning such content analyses within the larger conceptual frameworks of curriculum studies and the research and development of learning trajectories; and (3) reviewing recent high-profile content analyses that illustrate the need for more descriptive studies based on examining the learning progressions that are implied or defined by textbook authors. Then, I present a framework for ALT-based mathematical content analyses, outline the methodology used in a recent ALT-based analysis of four middle-school textbook series, and present selected results from the ALT-based analysis. In the final section, I discuss implications for the field with respect to utilizing the ALT construct to conduct informative ALT-based mathematical content analyses.

THEORETICAL CONSIDERATIONS

Articulated Learning Trajectories

To better understand key differences and similarities in the development and presentation of mathematical content among textbooks currently used by teachers, I conducted a mathematical content analysis of four middle-school textbook series, using the construct of an ALT as a lens to view the use of patterning concepts in the development of algebraic thinking constructs within the four series. A key purpose and outcome of the study, along with the identification of important differences among textbook series, was the development of an ALT-based framework to guide future mathematical content analyses. Although ALTs are not visible to teachers, developing the ALTs that articulate what can be inferred from mathematics textbook content as the learning progression in the content has the potential to inform teachers' decisions with respect to the mathematics content they present in their classrooms.

To begin, an ALT is defined as follows:

> An ALT is a learning trajectory that goes beyond the situated predictions of student learning associated with an HLT [hypothetical learning trajectory] (Simon, 1995). Instead, an ALT is a precise written sequence of mathematics content that authors of textbooks envision teachers and students will follow or use to guide the development of mathematics content within the classroom....[It] is a baseline of how a concept develops in written text. Certainly, the interaction of teachers and students with the written word will modify the trajectory; and teachers' development of ad hoc HLTs during lesson planning or classroom interactions will possibly substantially modify the use of the articulated learning trajectory. Nonetheless,...the ALT is a strong determining factor in the way the material will be presented in the classroom and interpreted during individual study from the textbook. (Olson, 2010, p. 68)

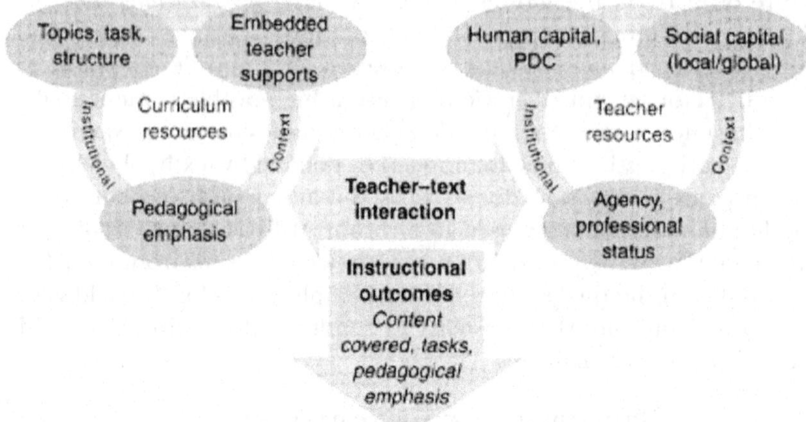

FIGURE 7.1. Teachers and curriculum. *Note:* Conceptual model of the relationships and interactions between teachers and curriculum presented by Remillard (2009, p. 89).

The ALT construct was developed with respect to mathematics curriculum and drew heavily on the research and discussions on *hypothetical learning trajectories* (Baroody, Cibulskiks, Lai, & Li, 2004; Battista, 2004; Clements, 2002; Clements & Sarama, 2004; Clements, Wilson & Sarama, 2004; Gravemeijer, 2004; Lesh & Yoon, 2004; Simon, 1995; Simon & Tzur, 2004; Steffe, 2004), design experiments (Brown, 1992; Cobb et al., 2003; Lehrer, Kim, & Schauble, 2007), and recent studies to elucidate students' development of mathematical knowledge utilizing learning trajectory constructs (Confrey, Maloney, Nguyen, Mojica, & Myers, 2009).

Conceptual Frameworks

With respect to broader frameworks for examining the role of textbooks in teaching and learning mathematics in classrooms settings, I specifically position this work within two recent conceptual models. In the first model (Figure 7.1), topics, tasks, and structure (aspects that I argue are related to ALTs in textbooks) are situated within the realm of curriculum resources, which, through instruction, pedagogical emphasis, teacher supports, and context, interact with teacher resources to produce instructional outcomes. Through this model, the relationship between ALTs and instructional outcomes is mediated through the interactions between teachers and textbooks.

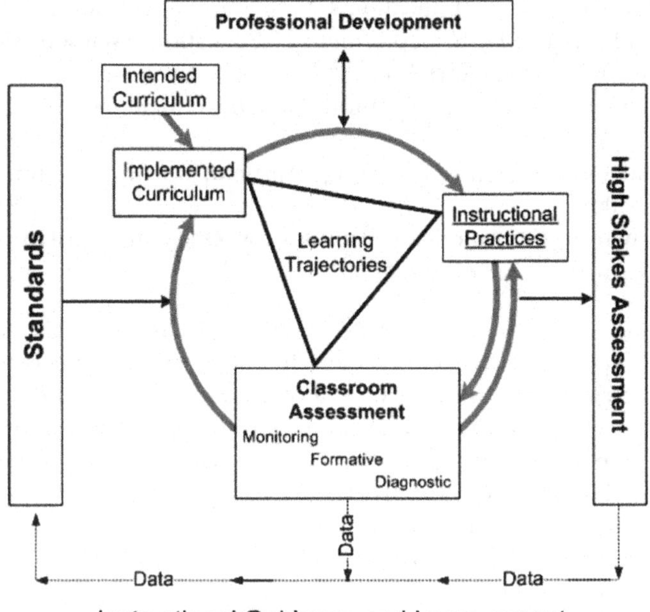

FIGURE 7.2. Embedding the instructional core in an accountability framework (Confrey, 2011)

In the second model (Figure 7.2), learning trajectories occupy a central role in connecting the implemented curriculum, instructional practices, and classroom assessment. In the model in Figure 7.1, the learning trajectories (i.e., the ALTs) present in the curriculum resources are likely related to the learning trajectories interacting between the intended curriculum and implemented curriculum in Figure 7.2, which are connected to the learning trajectories central to that model.

Content Analyses: Evaluative and Descriptive

Evaluation studies of curriculum materials have led to several high-profile content analyses that sought to determine the relative "worth" of curricula (see National Research Council, 2004; Stein, Remillard, & Smith, 2007). Five studies in particular are often identified (National Research Council, 2004; Olson, 2010; Stein et al., 2007):

- AAAS Project 2061 (American Association for the Advancement of Science, 2000)

- U.S. Department of Education's Exemplary and Promising Mathematics Programs (U.S. Department of Education's Mathematics and Science Expert Panel, 1999)
- Mathematically Correct (Clopton, McKeown, McKeown, & Clopton, 1999)
- the Adams report (Adams, Tung, Warfield, Knaub, Mudavanhu, & Yong, 2000)
- a study from the American Institutes for Research (Ginsberg, Leinwand, Anstrom, & Pollock, 2005)

The authors of these studies either compared curricula against each other with respect to evaluation of the value of particular aspects or evaluated the curricula against a set of external criteria (e.g., standards and benchmarks). However, the findings in each of these studies can often be condensed into a paradigm of "better/worse," "strong/weak," "adequate/deficient," and so forth (Olson, 2010).

To avoid such dichotomous appraisals of mathematics textbooks, I focused on describing the development of key mathematical ideas and the textual placement of those mathematical ideas. The intent of my study was to reduce the value-laden and evaluative language in favor of descriptions of the mathematical content of textbooks. Per the recommendations of the National Research Council (NRC) (2004) report *On Evaluating Curricular Effectiveness*, I oriented my study within the disciplinary perspectives of clarity, comprehensiveness, accuracy, depth of mathematical inquiry, and mathematical reasoning, organization, and balance.

Content analysis of textbooks was conducted to identify textual placement of mathematical concepts. ALTs were then formulated to describe the development of the concepts within an individual textbook or a textbook series. The ALTs were then scrutinized using the lenses of the six disciplinary perspectives recommended in the NRC report (2004). In the following section I provide a brief report on the mathematical content analysis I conducted using the ALT construct.

AN ALT-BASED MATHEMATICAL CONTENT ANALYSIS OF FOUR MIDDLE-SCHOOL TEXTBOOK SERIES

This study examined the way in which authors of four textbook series[2] utilize patterning concepts in the development of algebraic thinking constructs. In other words, an ALT-based mathematical content analysis was conducted of four textbook series to identify ALTs related to patterning concepts and algebraic thinking. The four textbook series examined were *Connected Mathematics 2* (*CMP2*) (Lappan, Fey, Fitzgerald, Friel, & Philips, 2009),3 *Mathematics: Applications and Concepts* (*Glencoe*) (Bailey et al., 2006),

MathThematics (Billstein & Williamson, 2008b), and *Saxon Math* (*Saxon*) (Hake, 2007).

The methodology consisted of several steps. The initial steps involved selecting an area of mathematical content, as well as the related grade-level span on which to focus the analysis. A coding scheme was developed, and instances of the selected mathematical content were identified within each of the textbooks. The instances were coded according to the coding scheme, an analysis was conducted using the coded instances, and related content was identified within textbooks that allowed for ALTs to be identified.

The following sections describe the methodology employed; however, the methodology outlined here can serve as a general framework for ALT-based mathematical content analyses in which ALTs can be identified, described, and established within textbooks and across textbook series.

Mathematical Content and Span of Curricula Studied

Determinations of content and grade-level span were informed by a review of literature on students' development of knowledge in the particular mathematical area, mathematical writings in relation to the particular topics, historical analyses of mathematics textbooks, and relevant standards documents: the *Common Core State Standards for Mathematics* (National Governors Association Center for Best Practices and Council of Chief State School Officers, 2010), *Principles and Standards for School Mathematics* (2000) of the National Council of Teachers of Mathematics (NCTM), and the NCTM's *Curriculum and Evaluation Standards for School Mathematics* (1989). By examining these sources, particularly with respect to studies on the development of mathematical learning trajectories based on students' thinking and abilities, a picture emerged of key concepts in the development of mathematical constructs and grade levels that are assumed appropriate for development of the selected concepts. This study concentrated on the development of algebraic thinking[4] concepts through patterning constructs. The study of patterns is consistently identified as associated with the fundamental work of mathematicians (Fey, 1990; Steen, 1990). Moreover, the ability to interpret the patterns that we observe is often recognized as an integral facet of our own understanding of our world (Senechal, 1990). However, the study of patterns alone does not necessarily imply that algebraic thinking constructs can then be built upon patterning concepts.

To better understand connections between patterning concepts and algebraic constructs associated with pattern recognition, extension, and generalization, I examined research related to algebraic thinking with respect to students' abilities regarding their conceptual understanding, thinking processes, and reasoning related to generalizations (Friedlander, Hershkowitz & Arcavi, 1989; Lannin, 2003, 2005; Lannin, Barker & Townsend, 2006a, 2006b; Lannin, Townsend, Armer, Green, & Schneider, 2008; Sasman, Olivier, &

Linchevski, 1999; Swafford & Langrall, 2000; Taplin & Robertson, 1995). Similar research examined students' use of actions, signs, and symbolization as related to generalizing patterns, including notations for functions (Lannin et al., 2008; MacGregor & Stacey, 1992, 1993, 1997; National Council of Teachers of Mathematics, 1989, 2000; Radford, 1999, 2000; Radford, Bardini & Sabena, 2007; Sabena, Radford & Bardini, 2005; Saul, 2008). Overall, the research establishes the importance of students' investigating and extending patterns, as well as utilizing symbols and signs in generalizing these patterns to develop understandings related to variable and function concepts in which an unknown value is dependent on other known values.

Reports and standards documents also emphasize the importance of patterning concepts in the development of algebraic thinking. NRC's report *Adding it Up: Helping Children Learn Mathematics* notes, "Through an emphasis on generalization, justification, and prediction, students can learn to use and appreciate algebraic expressions as general statements" (National Research Council, 2001, p. 279). The NCTM standards (1989, 2000) clearly identify studying patterns as important to algebraic thinking concepts, saying, "The study of patterns in grades 5–8 builds on students' experiences in K–4 but shifts emphasis to an exploration of functions.... It begins in K–4, is extended and made more central in 5–8" (NCTM, 1989, p. 98). Moreover, "In grades 6–8 all students should understand patterns, relations, and functions: represent, analyze, and generalize a variety of patterns with tables, graphs, words, and, when possible, symbolic rules" (NCTM, 2000, p. 222). Because much of the discussion related to patterning and algebraic thinking revolved around students in the middle grades (6–8), I determined the curriculum spans for this study to be grades 6–8, as well as within each of the grade levels 6, 7, and 8.

The relation of patterning concepts to algebraic thinking was further informed by historical examination of patterning concepts in textbooks over the past century (Olson, Regis & Papick, 2010) as well as an analysis of the mathematical soundness of common presentations of pattern problems (Papick, Olson & Regis, 2010). These two studies further clarified distinctions among the terms *pattern, sequence, arithmetic sequence,* and *geometric sequence.* Specifically, from a historical perspective, arithmetic and geometric sequences have been a ubiquitous presence in textbooks throughout the past century (Collins, 1911; Smith, 1924; Wells, 1908).

From a mathematical perspective, Papick, Olson, and Regis (2010) make clear that the definition of a *sequence* is a function that maps the positive integers into any nonempty set, a key mathematical structure not always evident in textbook authors' presentations of pattern problems. In their analysis of pattern-type problems, Papick et al. identify underlying mathematical structural concerns in problems not clearly utilizing sequence. In particular, such issues arise with pattern problems—such as the list of numbers 3, 5, 7, ... —and the task prompt for students to determine numbers occurring later in this list. Specifically, they argue, "There are infinitely many functions *g* defined from the positive integers into the integers having the values, $g(1) = 3$, $g(2) = 5$, and $g(3) = 7$, and moreover there are infinitely many polynomial functions with these given values" (Papick et al., 2010, p. 31). In relying on pattern problems that are not often anchored in the mathematical structure of sequences, discussion of the underlying structure of problems such as "3, 5, 7, ..." was lacking in problems analyzed by Papick et al. In particular, one such extension discussion the authors identify related to this problem is that "the only polynomial functions *g* that have the values $g(1) = 3$, $g(2) = 5$, and $g(3) = 7$ are induced by polynomials of all degrees other than two" (p. 31). As such, the authors argue that without focusing on the mathematical structure of a sequence, students' focus often is on determining expected solutions, instead of analyzing under which circumstances such problems make mathematical sense.

Coding Scheme and Guiding Questions

Codes corresponding to categories of various patterning concepts are presented in Table 7.1, along with descriptions related to each code. The coding scheme in this table was developed for the purpose of categorizing, examining, and describing concepts related to patterning and algebraic thinking in order to identify ALTs within the textbooks analyzed. The primary basis for the development of the coding scheme was the identification of content topics from the review of the research-based sources described above.

A set of guiding questions, outlined in Table 7.2, was developed to guide the assignment of codes to content in the textbooks. Guiding questions specify the way in which particular blocks of content are sorted into different codes, a necessary step for the subsequent analyses of textbooks. These guiding questions and resolutions thereof reflect the hierarchies of content and textbook codes outlined in Table 7.1. For example, the *context* of a pattern is either numeric (NP) or geometric (GP). The classification of the patterning constructs is slightly more flexible or variable. Each instance must be classified as pattern recognition (PC-R) (the base code for this study). However, a particular instance may also be coded as pattern extension (PC-E) and/or pattern generalization (PC-G).

The Nature of Instances and Applying the Coding Scheme

Once the coding scheme was developed, instances in textbooks were identified using the basic code for this content, pattern recognition, as a guide to determine what textbook material should be analyzed in this study. In this study, an instance serves as a way to consistently group textbook content for subsequent coding and analysis. The methods for grouping and

TABLE 7.1. Textbook Codes Corresponding to Patterning Concepts and Algebraic Thinking Constructs

Concepts	Category	Code	Description
Patterning Constructs	Patterning Concepts – Recognition constructs	PC-R	Students are prompted to recognize aspects of a provided pattern.
	Patterning Concepts – Extension constructs	PC-E	Students are prompted to provide an extension or various extensions of a pattern, or to use extensions in solving related problems.
	Patterning Concepts – Generalization constructs	PC-G	Students are prompted to provide a general rule (explicit, recursive, or otherwise descriptive) for a pattern. This categorization includes potentially using the generalization in solving related problems.
Context of Pattern	Numeric Context (numerically defined pattern)	NP	The structure of the pattern is based solely on numbers, or numbers derived from a situation in which there is no connection to an underlying visual or geometric context.
	Geometric Context (geometrically defined pattern)	GP	The structure of the pattern is inseparably connected to an underlying visual or geometric context.
Sequence Concepts	General Sequence Concepts	SQ	The authors explicitly identify a pattern as a sequence, or otherwise explicitly use the term sequence.
	Arithmetic Sequence Concepts	ASQ	The authors explicitly identify a pattern or sequence as an arithmetic sequence, provide a common difference to structure the pattern, or otherwise explicitly use the term arithmetic sequence.
	Geometric Sequence Concepts	GSQ	The authors explicitly identify a pattern or sequence as a geometric sequence, provide a common ration to structure the pattern, or otherwise explicitly use the term geometric sequence.
Variable Concepts	Implicit Variable	IV	Students are prompted to generalize a pattern, but not explicitly prompted to use a variable; or, students are prompted to find missing values in a pattern.
	Explicit Variable	EV	Students are prompted to provide a description, extension, or generalization of a pattern explicitly using a variable or variables; or, the authors explicitly us a variable in the presentation of a problem.
Function Concepts	General Function Concepts	FN	The authors explicitly connect the structure of a pattern to concepts regarding functions, function rules, or inputs and outputs.

TABLE 7.2. Guiding Questions for Textbook Coding

Concepts (hierarchically ordered)	Guiding Question	Resolution of the Guiding Question using the Table of Codes
Patterning Constructs	Through which pattern constructs are the mathematical concepts presented?	Pattern recognition constructs: PC-R; and Pattern extension constructs: PC-E; and/or Pattern generalization constructs: PC-G. These codes are hierarchical. All patterning problems are coded PC-R, but may also involve pattern extension or generalization.
Context of the Pattern	In what context is the pattern presented?	Numeric context: NP, or Geometric context: GP. A pattern is presented either in a numeric or geometric context. A numeric context is devoid of visual or geometric connections to the pattern. If there are visual connections to the underlying structure of the pattern, then the pattern has a geometric context.
Sequence Concepts	Is the pattern identified as a sequence?	General sequence: SQ, and/or Arithmetic sequence: ASQ, and/or Geometric sequence: GSQ. All patterns defined as sequences are coded SQ unless otherwise explicitly identified as arithmetic and/or geometric by labeling, identification of common differences or ratios, or a prompt to differentiate or otherwise define characteristics of such sequences.
Variable Concepts	Are variables implicitly or explicitly incorporated?	If students are prompted to provide descriptions, identifications, extensions, or generalizations of patterns but not explicitly prompted to present these in the form of nth terms or other forms using variables, there variable concepts are implicit: IV. If students are prompted to find the nth term of a pattern or sequence, or otherwise represent their descriptions, extensions, and generalizations in the form of a variable, then variable concepts are explicit: EV.
Function Concepts	Are function concepts related to input and output values incorporated?	An explicit identification of input values that produce values explicitly identified to be outputs via the relationships defined by the structure of the pattern, or by a function rule, is related to functional notions of input and output: FN. For the purposes of this study, function concepts in the absence of patterning constructs are not considered FN.

coding textbook content outlined in Table 7.3 delineate the four types of instances used to categorize the textual material that was analyzed. As such, the content on a relevant page[5] was classified as belonging to one of four specific "instances": a student problem or prompt that is a super-problem with one or more sub-problems, a student problem with no sub-problems, an author example or set of examples, or other author text.

TABLE 7.3. Methods for Grouping Textbook Content

Organization of Instances	Coding Procedure	Final Grouping of Instances for Analysis
A student problem consisting of a super-problem or prompt with one or more sub-problems.	Record relevant aspects of the super-problem, initial prompt, and sub-problems individually.	Group the super-problem or prompt, and sub-problems. Label the grouping with all relevant codes.
A student problem with no sub-problems.	Record the single problem.	Group the single problem. Label the grouping with relevant codes.
An author example, or group of related examples.	Record each related example individually.	Group the related author examples. Label the grouping with all relevant codes.
Author text before, between, or after author examples or student problems. This text is related to defining, elaborating on, or clarifying terminology involving patterning concepts.	Code each relevant aspect of the text individually. That is, code each definition, elaboration, or clarification of individual terms or concept separately.	Group the author text. Label the grouping with all relevant codes.

Within the 12 textbooks, each page was examined to identify every instance related to patterning. After the instances were identified, an iterative process was conducted that consisted of piloting and revising the coding scheme based on reliability work[6] and finalizing the scheme for use in analysis. Partial interrater reliability was assessed through the coordinated coding of 72 total pages with a mathematician and a mathematics educator; both hold tenure-track academic positions at a university in a department of mathematics. Through the reliability process, the coding scheme used for the study was finalized, and the final determination was made for the categorization of instances.

Standardized Page Positions of Content Instances

Each instance was also assigned *standardized page position* (SPP) (Flanders, 1994), a mechanism for providing information on the relative position of coded material in textbooks. The SPP was calculated as follows:

> To determine an overall picture of where items are located in textbooks, all pages were counted in each text from the opening of the first chapter through the last chapter test. (Supplementary exercises at the end of texts, glossaries, appendices, answer pages, and indices were not included in this count.)... [A]n average page position was calculated by standardizing the position of the item in each text as a percent of the way through the text. (Flanders, 1994, p. 68)

For example, an instance with an SPP of .48 occurs almost one half of the way through the textbook.[7] The purpose of determining an SPP for each instance is to provide a basis and structure from which a mathematical landscape of the content being examined begins to emerge. By assigning each instance an SPP, analysis can be conducted related to what instances come before and after other instances, and one can speculate about potential developmental reasons for the particular sequencing of concepts. Furthermore, by assigning an SPP to each instance, SPP spans can be determined for particular content.

Specifically, an *SPP span* is defined as the textual distance from the first instance involving a particular mathematical concept or representation to the final instance of similar concepts or representations. For example, suppose the first instance related to sequences occurs in a textbook at an SPP of .006 (essentially the beginning of the textbook), and the final instance of such concepts occurs at an SPP of .98 (essentially at the end of the textbook). In such an example, the SPP span of sequence concepts across this textbook would be from .006 to .98. However, it is important to note that because sequence concepts might be found in varying densities throughout this hypothetical textbook, other representations, discussed later, were developed to understand the frequency, density, and order in which these instances occur across the textbook.

An SPP and SPP span provide different layers in understanding the mathematical landscape of textbook content. Specifically, I take the *mathematical landscape* of a textbook to be defined by the placement of the mathematical content within the textbook without necessary regard to any underlying conceptual connections among the content. That is, developing a working map of a mathematical landscape in a textbook allows the identification of certain key features of the landscape—a "mountain" of sequence concepts one fourth of the way through the textbook, or a span of variable concepts related to patterning across the middle third of the textbook. The landscape necessarily exists even without delineation of possible developmental connections among key features of the mathematical landscape of the textbook. In other words, such data as an SPP or SPP span provide key orienting landmarks, but by themselves do not necessarily depict connections among the features mapped out in the landscape.

Consequently, an SPP or SPP span does not necessarily determine or identify an ALT within a textbook. Rather, the representations and descriptions of SPP spans, individual SPP data points, and tallies of the number of instances assigned particular codes (i.e., code counts) provide orienting information about various aspects (or key features) of the textual mathematical landscape. From these key features of the landscape, further analysis of the relationships among particular instances and spans allows for the development and identification of ALTs within a textbook.

Compilation of Data, Code Counts, and Mapping SPP Spans and ALT Models

The information associated with each instance, including page numbers, codes, SPP numerals, and a narrative description of the nature of the content of each instance was catalogued in a spreadsheet. Code counts were then conducted across various combinations of codes to better understand the degree to which certain mathematical concepts arose in a given textbook. For example, a code count was conducted for instances involving pattern recognition constructs (i.e., all of the instances coded as PC-R, or PC-R and a combination of variable concepts, sequence concepts, and/or function concepts). Another code count was conducted of the instances involving pattern recognition constructs in geometric contexts (e.g., the instances coded PC-R, GP; or PC-R, GP and a combination of variable concepts, sequence concepts, and/or function concepts). Consequently, the number of instances involving pattern recognition constructs in numeric contexts could be determined by taking all of the instances coded as PC-R and subtracting all of the instances coded as PC-R, GP.

The purpose of conducting these code counts was to map key features of the mathematical landscape of the textbook. In this study, a map emerges of how many instances involving pattern recognition constructs are presented in a numeric context versus a geometric context. Although this information does not provide for the identification of an ALT, it does provide orienting information about the degree to which the textbook authors incorporate visual versus strictly numeric representations of patterns. Together with information regarding the SPPs and SPP spans of these instances, connections arise among these particular features of the landscape of the textbook, connections from which ALTs were identified and described.

Along with determining the number of instances assigned particular codes, SPP span and sequence graphs were developed to provide further visualization of the landscape of the textbook. From these SPP span and sequence graphs, and the analysis of instances through descriptions of the connections across the content in coded instances, working visual models of ALTs were developed. In the following section, selected results from code counts are provided, as are selected SPP span and sequence graphs and selected working visual models of ALTs.

SELECTED RESULTS FROM THE CONTENT ANALYSIS

The data presented in this section were gathered through the ALT-based mathematical content analysis of 12 middle-school textbooks (Grades 6, 7, and 8) from the four textbook series named in the previous section. A total of 7,573 textbook pages across the four textbook series were examined; following is a selection of results from the ALT-based analysis.

Three major results emerged from the ALT-based mathematical content analysis. First, inferences drawn with regard to the nature of the mathematics among the four textbook series were observed through examining code counts and SPP spans along with implications regarding the presentation, inclusions, and scope of mathematics in the textbooks. Second, ALTs were identified within each textbook (and textbook series) with respect to the presentation of mathematics, its scope and sequencing, and discernable connections that could be identified in the authors' presentation of the material within the textbooks and textbook series. The third result was a proof of concept. That is, identifying the number of instances of certain content, mapping the SPPs and SPP spans of those instances, and the resulting identification of ALTs within textbooks and textbook series demonstrated that the framework and methodology outlined in this chapter provided a well structured and consistent means of identifying the development of mathematic concepts within textbooks.

Instances Among the Textbook Series

In the 12 textbooks, a total of 1,032 instances of content were identified and coded (see Table 7.4 for instances by textbook series) relating to the patterning problem construct (PC-R, PC-E, and/or PC-G), the context of the patterns (NP or GP), and the inclusion of sequence type (SQ, ASQ, and/or GSQ), variable (EV or IV), and function (FN) concepts.

Tables were created for each of the 12 textbooks to display the number of code counts, the SPP span of the codes, and the percentages associated with total counts. No SPP span is reported for a textbook series because SPP spans are most meaningful within each textbook rather than across a textbook series.

In Table 7.5, *n* represents the number of instances coded with a combination of the column code and row code for a particular cell. The codes used in this table directly relate to the codes presented in Table 7.1. Additionally, codes with a "slash" indicate a dual count—SQ/FN are the instances involving both sequence and function concepts. The percentage represents the percentage of the total number of instances coded as the same combination with respect to the parenthetical cues in the totals column and row. For example, the shaded cells in the center of the table represent the number of coded instances involving any patterning constructs with function concepts that are presented in numeric or geometric contexts. Thus, in *CMP2,* 73% of instances involving function concepts and patterning constructs are presented in numeric contexts, and 27% of similar instances are presented in geometric contexts. A total of 102 instances involving patterns related to function concepts are presented in numeric contexts. Those 102 instances represent 54% of all of the instances that are presented in a numeric context and 73% of all of the instances involving function concepts. Similarly,

TABLE 7.4. Number of Instances of Patterns Per Textbook Series (Grades 6–8)

Textbook Series	# Instances (Gr. 6–8)
CMP2	271
Glencoe	245
MathThematics	227
Saxon	289

TABLE 7.5. Data Related to Patterning Constructs, Pattern Structures, and Algebraic Thinking Constructs in the CMP2 Textbook Series

Codes		PC-R	Up to PC-E	Up to PC-G	NP	GP	Totals
SQ	n	0	1	1	1	1	2
	%	0% (0%)	2% (50%)	0.7% (50%)	0.5% (50%)	1% (50%)	0.7% (100%)
ASQ/GSQ	n	0	0	0	0	0	0
	%	0% (0%)	0% (0%)	0% (0%)	0% (0%)	0% (0%)	0% (100%)
EV/IV	n	16	7	128	106	45	151
	%	23% (11%)	12% (5%)	90% (85%)	56% (70%)	56% (30%)	56% (100%)
FN	n	31	15	94	102	38	140
	%	44% (22%)	26% (11%)	66% (67%)	54% (73%)	47% (27%)	52% (100%)
SQ/V	n	0	0	1	0	1	1
	%	0% (0%)	0% (0%)	0.7% (100%)	0% (0%)	1% (100%)	0.4% (100%)
SQ/FN	n	0	0	1	0	1	1
	%	0% (0%)	0% (0%)	0.7% (100%)	0% (0%)	1% (100%)	0.4% (100%)
V/FN	n	13	3	92	78	30	108
	%	19% (12%)	5% (3%)	64% (85%)	41% (72%)	37% (28%)	40% (100%)
Totals	n	70	58	143	190	81	271
	%	100% (26%)	100% (21%)	100% (53%)	100% (70%)	100% (30%)	100% (100%)

a total of 38 instances involving patterns related to function concepts are presented in geometric contexts. Those 38 instances represent 47% of all of the instances that are presented in a geometric context and 27% of all of the instances involving function concepts.

The data in Table 7.6 indicate that 90% of function concepts involving patterns in *Glencoe* are presented in a numeric context, and 10% of similar concepts are presented in a geometric context. Given that 73% of instances involving function concepts in *CMP2* are presented in numeric contexts and 27% are presented in geometric contexts, a student studying from *Glencoe* will encounter much fewer instances in which function concepts are presented in geometric contexts.

TABLE 7.6. Data Related to Patterning Constructs, Pattern Structures, and Algebraic Thinking Constructs in the Glencoe Textbook Series

Codes		PC-R	Up to PC-E	Up to PC-G	NP	GP	Totals
SQ	n	16	23	3	47	5	52
	%	31% (31%)	17% (44%)	5% (6%)	24% (90%)	10% (10%)	21% (100%)
ASQ/ GSQ	n	13	20	1	32	2	34
	%	25% (38%)	14% (59%)	2% (3%)	16% (94%)	4% (6%)	14% (100%)
EV/IV	n	16	9	51	68	8	76
	%	31% (21%)	8% (12%)	85% (67%)	35% (89%)	16% (11%)	31% (100%)
FN	n	13	11	38	56	6	62
	%	25% (21%)	9% (18%)	63% (61%)	29% (90%)	12% (10%)	25% (100%)
SQ/V	n	7	4	3	13	1	14
	%	14% (50%)	3% (29%)	5% (21%)	7% (93%)	2% (7%)	6% (100%)
SQ/FN	n	0	1	1	1	1	2
	%	0% (0%)	1% (50%)	2% (50%)	1% (50%)	2% (50%)	1% (100%)
V/FN	n	6	5	37	44	5	48
	%	12% (13%)	4% (10%)	62% (77%)	23% (92%)	10% (8%)	20% (100%)
Totals	n	51	134	60	194	51	245
	%	100% (21%)	100% (55%)	100% (24%)	100% (79%)	100% (21%)	100% (100%)

In addition, in Table 7.6, 21% of all instances related to patterning concepts within the *Glencoe* textbook series involve patterns being identified as sequences (as identified by the nonparenthetical percentage in the SQ-Totals cell). Furthermore, of all the instances involving sequences (i.e., patterns explicitly identified as being a sequence), 90% are presented within a numeric context, and 10% are presented within a geometric context (as identified by the parenthetical percentages in the SQ-NP and SQ-GP cells, respectively).

Tables 7.5, 7.7, and 7.8 present analogous data for the *CMP2, MathThematics* and *Saxon* series, respectively. In particular, there were only two instances in *CMP2* in which a pattern was explicitly identified as being a sequence, which amounts to only 0.7% of all of the instances of patterning in *CMP2*. In *Saxon*, 45% of patterns are identified as being a sequence. Similarly, in *MathThematics*, 40% of patterns are identified as being sequences. These

TABLE 7.7. Data Related to Patterning Constructs, Pattern Structures, and Algebraic Thinking Constructs in the MathThematics Textbook Series

Codes		PC-R	Up to PC-E	Up to PC-G	NP	GP	Totals
SQ	n	13	36	41	62	28	90
	%	46% (14%)	35% (40%)	43% (46%)	38% (69%)	44% (31%)	40% (100%)
ASQ/ GSQ	n	8	1	6	12	3	15
	%	29% (53%)	1% (7%)	6% (40%)	7% (80%)	5% (20%)	7% (100%)
EV/IV	n	5	6	66	59	18	77
	%	18% (6%)	6% (8%)	69% (86%)	36% (77%)	28% (23%)	34% (100%)
FN	n	2	8	40	36	14	50
	%	7% (4%)	8% (16%)	42% (80%)	22% (72%)	22% (28%)	22% (100%)
SQ/V	n	4	3	36	32	11	43
	%	14% (9%)	3% (7%)	38% (84%)	20% (74%)	17% (26%)	19% (100%)
SQ/FN	n	0	0	18	9	9	18
	%	0% (0%)	0% (0%)	19% (100%)	6% (50%)	14% (50%)	8% (100%)
V/FN	n	0	1	33	23	11	34
	%	0% (0%)	1% (3%)	34% (97%)	14% (68%)	17% (32%)	15% (100%)
Totals	n	28	103	96	163	64	227
	%	100% (12%)	100% (45%)	100% (42%)	100% (72%)	100% (28%)	100% (100%)

percentages do not necessarily indicate that the lack—or prevalence—of identifying a pattern as a sequence is inherently beneficial, but it is an important consideration in analyzing the way in which authors build upon sequence definitions and concepts. Students studying from *CMP2* may not have an abundance of exposure to the term *sequence* from a curricular perspective; however, students studying from *CMP2* are also not exposed to textbook in which nearly one half the instances involve the referent of the term *sequence* with regard to a pattern.

Furthermore, in *Saxon*, 97% of the instances involving sequence concepts are presented in a numeric context. By comparison, sequence concepts presented in numeric contexts in *MathThematics* accounted for 69% of all of the instances involving sequences. In other words, a student studying from the *Saxon* textbook series will see fewer instances in which sequence

TABLE 7.8. Data Related to Patterning Constructs, Pattern Structures, and Algebraic Thinking Constructs in the Saxon Textbook Series

Codes		PC-R	Up to PC-E	Up to PC-G	NP	GP	Totals
SQ	n	24	74	32	126	4	130
	%	40% (18%)	58% (57%)	32% (25%)	56% (97%)	17% (3%)	45% (100%)
ASQ/ GSQ	n	7	2	4	13	0	13
	%	12% (54%)	2% (12%)	4% (31%)	6% (100%)	0%	4% (100%)
EV/IV	n	16	18	71	100	5	105
	%	27% (15%)	14% (17%)	70% (68%)	44% (95%)	22% (5%)	36% (100%)
FN	n	35	22	64	113	8	121
	%	58% (29%)	17% (18%)	63% (53%)	50% (93%)	35% (7%)	42% (100%)
SQ/V	n	3	10	15	27	1	28
	%	5% (11%)	8% (36%)	15% (54%)	12% (96%)	4% (4%)	10% (100%)
SQ/FN	n	0	0	1	1	0	1
	%	0% (0%)	0% (0%)	1% (100%)	0.4% (100%)	0% (0%)	0.3% (100%)
V/FN	n	12	5	55	67	5	72
	%	20% (17%)	4% (7%)	54% (76%)	30% (93%)	22% (7%)	25% (100%)
Totals	n	60	128	101	266	23	289
	%	100% (21%)	100% (44%)	100% (35%)	100% (92%)	100% (8%)	100% (100%)

concepts are presented in geometric contexts than if they had studied from the *MathThematics* textbook series.

With respect to the context in which patterns are presented among the four series, a geometric context is provided in *Glencoe* for 21% of the instances, in *CMP2* for 30% of the instances, in *Saxon* for 8% of the instances, and in *MathThematics* for 28% of the instances. Although these percentages in and of themselves do not describe a discernable ALT, they are examples of key features used to map the mathematical landscape of these textbook series. In particular, from these percentages, it could be expected that in general, any ALT identified in a *CMP2* textbook would include more instances involving geometric contexts than an ALT identified in one of the *Saxon* textbooks.

The data in Tables 7.5–7.8 also indicate differences in the instances that involve function concepts in the four series. Specifically, across each of the four textbook series, the percentage of instances related to patterning that involved function concepts is as follows: *Saxon*, 42%; *Glencoe*, 25%; *MathThematics*, 22%; *Connected Mathematics*, 52%. As with the other data in these tables, although the data do not specifically delineate ALTs, the counts illustrate that mathematical content often cited as important to the development of algebraic thinking, with respect to patterning concepts, varies widely with respect to its inclusion or exclusion from the various curricula. This variation does not necessarily imply inherent quality or lack of quality. These data suggest that regarding patterning concepts, connections between patterning constructs and function concepts are made in over one half the instances throughout the three textbooks of the *CMP2* series. Alternatively, in *Glencoe*, pattern constructs are connected to function concepts in one fourth of the instances throughout the three textbooks in the series. That is, when engaging in patterning concepts, students studying from *CMP2* will encounter a higher percentage of instances, relative to the overall number of instances involving patterning within the three textbooks of the series, in which function concepts are presented than will students studying from *Glencoe*.

Graphic Representations of SPP Spans: Toward Visual Representations of ALTs

The graphic representations of spans and sequences of the SPPs of instances across textbooks provided a further interpretation of data and more landmarks in the developing map of key features in the various mathematical landscapes. Figures 7.3 through 7.6 illustrate the progression of instances involving patterning constructs (PC-R; PC-R, PC-E; PC-R, PC-G; and PC-R, PC-E, PC-G) with respect to context (NP or GP) across the four grade 6 textbooks. In this regard, these figures represent the density of mathematical content within these SPP spans.

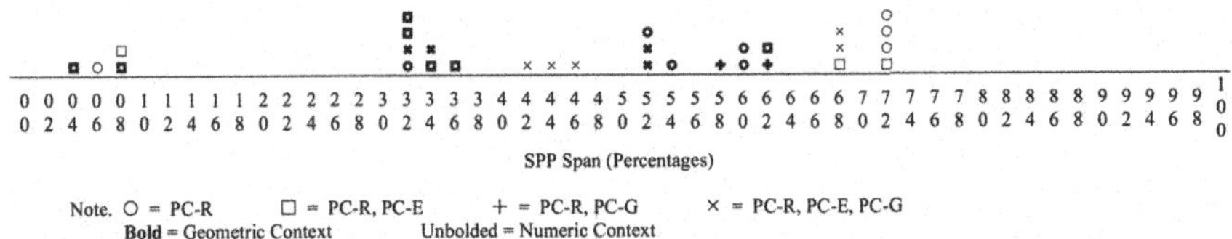

FIGURE 7.3. SPP span and patterning, *CMP2*, Grade 6. *Note:* Graph of SPP span and sequence of instances of patterning in *CMP2*, Grade 6. Closed-figure symbols (circle or a square) and open-figure symbols ("x" or "+") are utilized to distinguish between instances not involving pattern *generalization* constructs (PC-R; and PC-R, PC-E) and those involving pattern *generalization* constructs (PC-R, PC-G; and PC-R, PC-E, PC-G). Differences within the closed and open-figure symbols (i.e., between circle and square, and "x" and "+") indicate the inclusion of pattern *extension* constructs. An unbolded symbol indicates an instance involving a *numeric* context; a bolded symbol indicates an instance involving a *geometric* context. The vertically paired numbers below the horizontal line indicate two-place decimal numbers that refer to a relative SPP value. Symbols representing instances of patterning concepts are placed above the line to indicate the SPP interval within which the instance occurs, such that the interval within which an instance occurs is defined by the two-place decimal number to the left (exclusive lower bound of the interval) and the number directly under the symbol (inclusive upper bound). For example, the first instance in *CMP2*, grade 6, involving pattern recognition and extension constructs in a geometric context, appears at an SPP of .040, so a bolded square appears above the vertical 04, and lies within the interval (02, 04]. Similarly, the final instance in this textbook, involving pattern recognition constructs in a numeric context, in this textbook, occurs at an SPP of .719, so an unbolded circle was placed above the vertical 72, because that instance occurs within the interval (70, 72]. Multiple instances within the same interval are stacked vertically above the number representing the upper boundary of the interval, in ascending order according to their relative SPP positions. See text for additional explanation.

In each of these SPP span and sequencing graphs, the numbers below the line represent the SPP associated with each instance. For example, "14" printed vertically corresponds to an SPP of .14, which covers the SPP span from .121 to .14, or alternatively, corresponds to the interval between 12 (exclusive) and 14 (inclusive)—that is, (12, 14]. In other words, if an instance has an SPP of .138, it was graphed above the number 14. The stacks of symbols above the line on the graph represent the ascending sequence within the particular SPP span. The lowest symbols in the stack represent the values closest to the lower bound (e.g., .121); the uppermost symbols represent the values closest to the upper bound (e.g., .14).

The frequency with which students encounter patterns in numeric contexts versus geometric contexts varies as well, as evidenced by the presence of bolded versus unbolded symbols. That is, a bolded symbol represents an

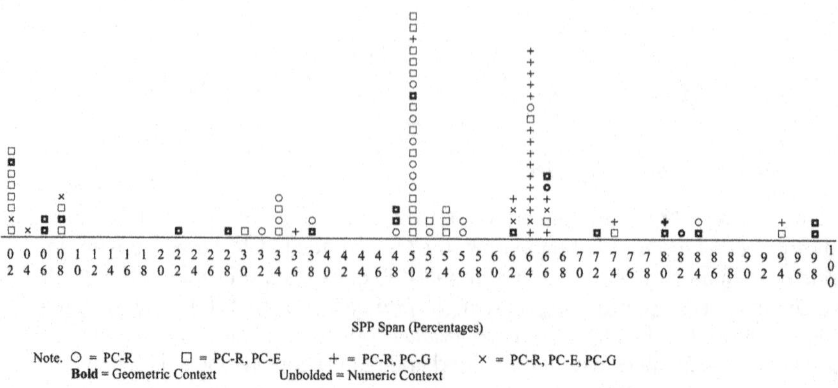

FIGURE 7.4. SPP span and patterning, *Glencoe Course 1,* Grade 6. *Note:* Graph of SPP span and sequence of instances of patterning in *Glencoe, Course 1,* Grade 6. For description of symbol notation, see note for Figure 7.3. Note that in *Glencoe, Course 1,* Grade 6, the first instance, involving pattern recognition and extension constructs in a numeric context, occurs at an SPP of .009, and as such appears as an unbolded square over the 02 since it is within the interval (0, 02]. The final instance, involving pattern recognition and extension constructs in a geometric context, in this textbook occurs at an SPP of .976, and as such appears as a bolded square over the 98 since it is within the interval (96, 98].

instance in which patterns are presented in a geometric context, and an unbolded symbol represents an instance in which patterns are presented solely in a numeric context.

A final aspect of these graphic representations is the type of symbols used in the representation. Open-figure symbols ("x" or "+") represent instances involving pattern generalization constructs, whereas closed-figure symbols (circle or a square) represent instances that do *not* involve pattern generalization constructs. That is, the circle and square represent the codes PC-R; and PC-R, PC-E, respectively. The "x" and "+" represent the codes PC-R, PC-G; and PC-R, PC-E, PC-G, respectively. As with other representation of other components of the content analysis, there are notable differences in the SPP densities, sequences, and nature of patterning problems among the textbooks across the four series.

In these four graphs of SPP span and sequence, a mathematical landscape was defined by delineating the locations within a textbook that students could encounter patterning problems. It is evident that the number of times students encounter patterning, as well as the location or timing of students' encounters with such content, varies widely among the four grade 6 textbooks represented in Figures 7.3 through 7.6.

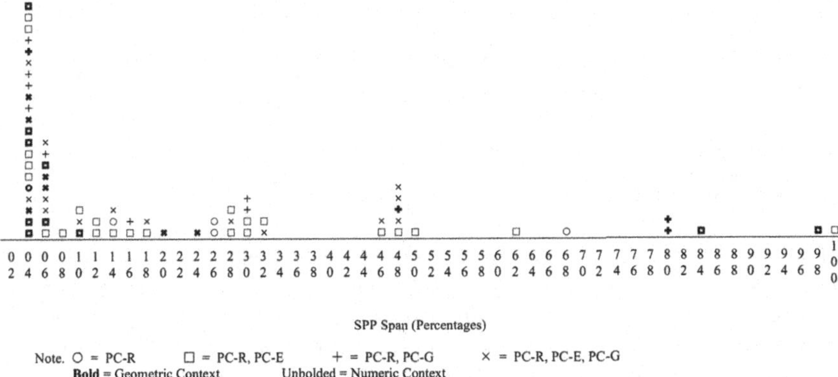

FIGURE 7.5. SPP span and patterning in *MathThematics,* Book 1, Grade 6
Note: Graph of SPP span and sequence of instances of patterning in *MathThematics,* Book 1, Grade 6. For description of symbol notation, see note for Figure 7.3. Note that in *MathThematics,* Book 1, the first instance, involving pattern recognition and extension constructs in a geometric context, occurs at an SPP of .027, and as such appears as a bolded square over the 04 since it is within the interval (02, 04]. Also note that there are 21 instances of patterning that occur within the SPP span of .027 to .039, represented by the 21 symbols within the interval (02, 04], 10 of which are presented with a geometric context (as denoted by the bolded symbols) and 11 of which are presented with a numeric context (as denoted by the unbolded symbols). The final instance, involving pattern recognition and extension constructs in a numeric context, occurs at an SPP of .986, and as such appears as an unbolded square over the 100 since it is within the interval (98, 100].

Other notable features of the mathematical landscape of the textbooks that arise from these graphs are related to key mathematical ideas that are clustered or that appear only sparingly. For example, in Figure 7.5, a dense concentration of "crosses" appears at the front of the textbook in *MathThematics,* indicating 17 instances involving pattern generalization constructs within the first one tenth, (00, 10], of the grade 6 textbook. A secondary minor presence of five instances involving generalization appears again just before one half of the way through the textbook, within the interval (46, 48]. Conversely, as seen in Figure 7.4, a concentration of 21 instances involving pattern generalization does not appear until three fifths of the way through the *Glencoe* textbook, within the interval (62, 66]. There also appears to be a secondary minor presence of three instances of pattern generalization within the interval (00, 08] in *Glencoe,* and only five other instances of pattern generalization exist in grade 6.

The implied exposure to pattern generalization concepts here is that students studying from these two grade 6 textbooks will have, essentially,

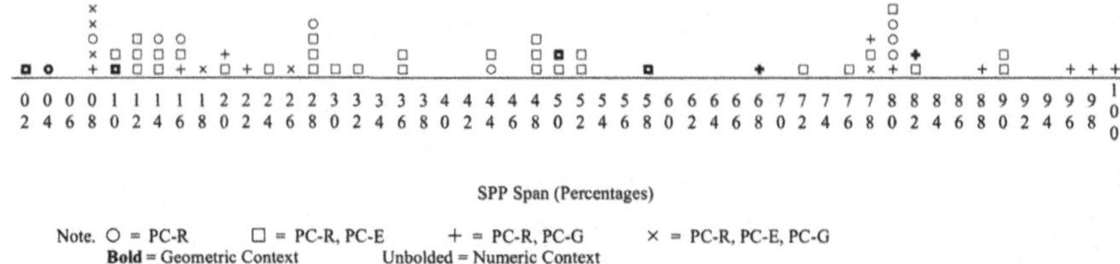

FIGURE 7.6. SPP span and patterning in *Saxon, Course 1,* Grade 6. *Note:* Graph of SPP span and sequence of instances of patterning in *Saxon, Course 1,* Grade 6. Note that in *Saxon, Course 1,* Grade 6, the first instance, involving pattern recognition and extension constructs in a geometric context, occurs at an SPP of .002, and as such appears as a bolded square over the 02 since it is within the interval (0, 02]. The final instance in this textbook, involving pattern recognition and generalization constructs in a numeric context, occurs at an SPP of .983, and as such appears as an unbolded "+" over the 100 since it is within the interval (98, 100].

inverted experiences. That is, a student studying from *MathThematics* from front to back will receive early and immediate exposure to pattern generalization constructs within a curriculum span that also includes patterns presented in numeric contexts and patterns presented in geometric contexts in a nearly 50/50 mix. This same student will then revisit pattern generalization constructs a little before one half of the way through the textbook. Alternatively, a student studying from *Glencoe* will receive limited exposure to pattern generalization constructs within an initial curriculum span in which patterns are dominantly presented in numeric contexts. This same student will then receive more intense exposure to pattern generalization constructs a little over three fifths of the way through the textbook, but these instances of patterning will once again be framed primarily by numeric contexts. In fact, 24 of the 27 instances in the interval (62, 66] are presented in numeric contexts, including every instance involving pattern generalization constructs. Such differences are precisely of the kind that must be acknowledged and understood in order to interpret the landscape of the textbook and to identify and describe any potential ALTs within the textbooks.

Graphic Representations of ALTs: Working Visual Models Based on SPP Span Visuals

The graphic representations of SPP span and sequence discussed in the previous section allowed for examinations and investigations of the mathematical landscape of the textbook.[8] Working visual models for ALTs based on qualitative narratives and utilizing the graphs of SPP span and sequence and other data presented in this chapter provide another representation, a general map, of the scope and development of content within textbooks.

The process used to develop working visual models of ALTs consisted of three steps. First, the graph of SPP span and sequence was examined in relation to the relevant content (e.g., all instances of patterning or only the instances of patterning involving sequence concepts). Second, an initial trajectory outline was developed with key aspects of the mathematical landscape of the textbook identified, including SPPs and related codes. Third, a descriptive working visual model was developed with respect to the first two steps and qualitative descriptions developed during the ALT-based content analysis.

Working Visual Model of an Identified Patterning Concepts—ALT Across Glencoe, Course 1, Grade 6

Figure 7.4 shows the graph of SPP span and sequence related to patterning concepts in *Glencoe*, grade 6. From this graph, and from other data collected throughout the study, the initial trajectory outline in Figure 7.7 was developed. In Figure 7.7, key points in which the mathematical focus of instances appears to change are identified across the textbook. In particular, one key instance identified as a "bridging and table problem" occurs at an SPP of .538. The significance of this problem in which the pattern is in a numeric context and presented in a table format is that the instances that precede it are all similar with respect to pattern extension constructs, but subsequent instances are similar with respect to the inclusion of pattern extension *and* pattern generalization constructs and function concepts presented in table formats (e.g., *x*, *y*—or other variable pairs—tables in which students are prompted to organize input and output values). However, this instance occurs within a lesson on problem-solving strategy and is titled "standardized test practice." Although this instance is one that appears to bridge extension constructs, and extension and generalization constructs within the context of a table (which itself implicitly connotes the concept of function), it occurs within a lesson of the textbook that potentially is not highlighted during the course of instruction. That is, although this problem is mathematically and structurally significant, the placement of the problem may not indicate such significance. Such instances that provide content connections or bridges are potential critical features of ALTs, and can perhaps serve as pivotal points for instructional focus. These instances may also serve as focal points in analyzing enactment of curricula as related to identified ALTs.

From this initial trajectory outline, as well as qualitative descriptions developed during that study, a working visual model of an ALT in *Glencoe, Course 1* can be identified. In particular, the narratives relevant to the trajectory outlined in Figure 7.7 indicated this ALT to be "related to function

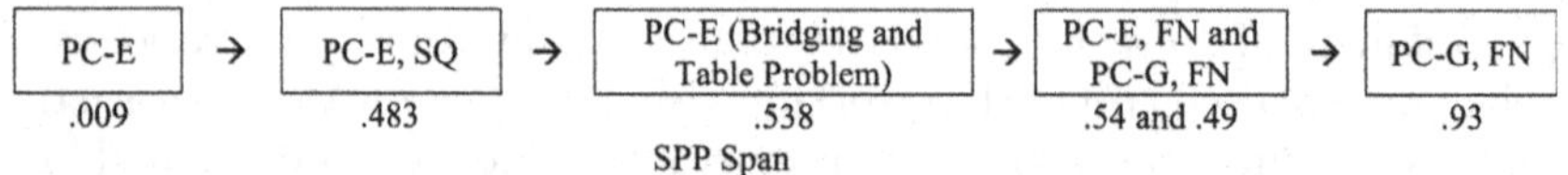

FIGURE 7.7. Trajectory graph of ALT for *Glencoe, Course 1,* Grade 6. *Note:* Initial trajectory graph of an ALT spanning *Glencoe, Course 1,* Grade 6. The codes in the boxes represent content as delineated by the coding scheme presented in Table 7.1. Other notes are provided for purposes of identifying key features of the instances within this potential trajectory. For example, "Bridging and Table Problem" refers to an instance in the "test practice section" of the textbook at an SPP of .538 related to instances of patterning that preceded it, but it was presented in a table format similar to instances of patterning involving function concepts that occur after.

concepts...developed with a pattern extension problem that...is structurally and conceptually consistent with...previous...pattern extension constructs.... [The subsequent] pattern generalizations involving function and variable concepts is [*sic*] based on [the] prior concepts related to...pattern extension constructs" (Olson, 2010, pp. 187–188). Figure 7.8 shows a working visual model for an ALT that spans the *Glencoe, Course 1* textbooks. In this working visual model, arrows represent threads of content with particular foci. The left and right ends of the figure represent the beginning and end of the textbook. Arrows branching from other arrows represent approximate positions in the textbook span at which discernable changes in content, comprising a new thread of content, occurred. SPP spans are included in the descriptive phrasings to provide reference for the approximate length of arrows, as well as indicate when in the textbook one should be able to find this content and the respective changes in the focus and structure of patterning constructs.

Working Visual Model of an Identified Patterning Concepts–ALT Across CMP2, Grade 8

A working visual model for a *textbook-spanning ALT* within *CMP2,* grade 8 was developed using processes similar to those used for the model in Figure 7.8. The SPP span and sequence of instances of patterning in *CMP2,* grade 8 are graphed in Figure 7.9. Clusters of content across four general intervals—(00, 10], (20, 32], (32, 44], and (70, 74]—correspond, respectively, to the following sections of the *CMP2,* grade 8 textbook: "Thinking with Mathematical Models," "Growing, Growing, Growing," "Frogs, Fleas, and Painted Cubes," and "The Shapes of Algebra." Taken together, these four content clusters form the textbook-spanning ALT in *CMP2,* grade 8.

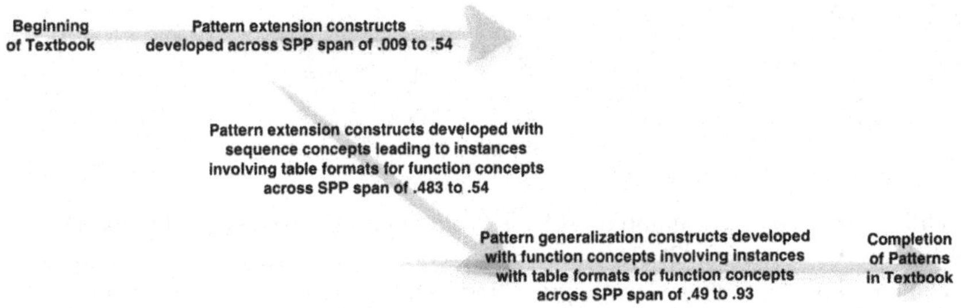

Beginning of Textbook

Pattern extension constructs developed across SPP span of .009 to .54

Pattern extension constructs developed with sequence concepts leading to instances involving table formats for function concepts across SPP span of .483 to .54

Pattern generalization constructs developed with function concepts involving instances with table formats for function concepts across SPP span of .49 to .93

Completion of Patterns in Textbook

FIGURE 7.8. Visual model for ALT for *Glencoe, Course 1*. *Note:* A visual model for an identified ALT spanning *Glencoe, Course 1,* Grade 6. The three arrows represent three overall content trends within this ALT. The descriptions are provided to highlight the nature of the content. The related curriculum SPP spans are included to indicate in which part of the textbook one should expect to find the concepts presented in the model.

Through examination of the data in Figure 7.9, as well as identification of relevant intervals throughout this textbook, an initial textbook-spanning ALT graph was developed, as shown in Figure 7.10. Qualitative descriptions formulated during this study were also examined in order to identify any subsequent ALTs within the textbook-spanning ALT in the *CMP2* grade 8 textbook. Of note, a *functional relationships ALT,* "related to the development of algebraic thinking constructs as embodied by variable and function concepts with respect to linear, exponential, and quadratic relationships" was identified (Olson, 2010, p. 285). This ALT spans the first three boxes

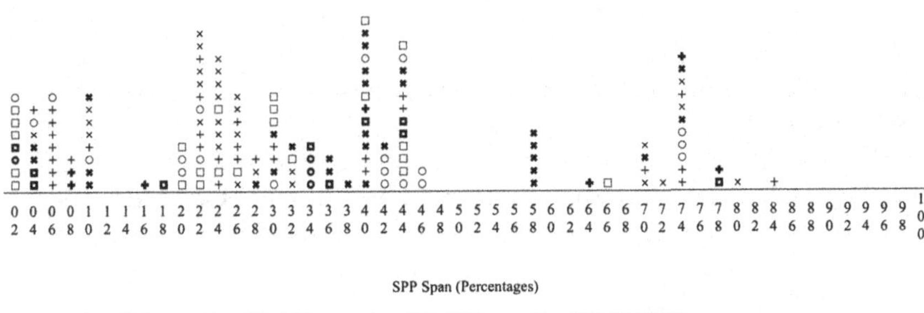

FIGURE 7.9. SPP span and patterning in *CMP2*, Grade 8
Note: Graph of SPP span and sequence of instances of patterning in *CMP2*, Grade 8. For descriptions of the bolding of symbols, and other utilization of symbols, see notes for Figures 7.3, 7.4, 7.5, and 7.6.

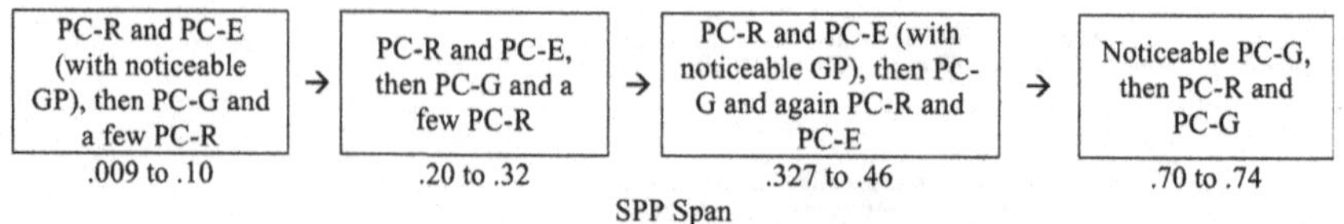

FIGURE 7.10. Trajectory graph of ALT for *CMP2,* Grade 8. *Note:* Initial trajectory graph of an ALT spanning *CMP2,* Grade 8. The codes in the boxes represent content as delineated by the coding scheme presented in Table 7.1. Other notes are provided for purposes of identifying key features of the instances within this potential trajectory.

in Figure 7.10. The first box pertains to the development of linear function concepts, the second box to the development of exponential function concepts, and the third box to the development of quadratic function concepts.

Within the functional relationships ALT, *underlying ALTs* were identified within each of the boxes associated with the development of each of the functional relationships that comprise the functional relationships ALT (Olson, 2010). In particular, the underlying ALTs of each of the function families within the larger functional relationships ALT initially generally include instances with "situational contexts that go beyond simple numeric or geometric contexts. These initial instances often do not involve pattern generalization concepts; instead, the instances generally involve the development of variable and function concepts through pattern recognition or pattern extension constructs" (Olson, 2010, p. 285). After the pattern recognition and extension constructs are initially developed, pattern generalization constructs are integrated with respect to variable and function concepts. Lastly, within the underlying ALTs that are positioned within the functional relationships ALT, instances are incorporated that involve more procedural, less contextual pattern recognition constructs in numeric contexts (Olson, 2010).

Lastly, a final aspect of the textbook-spanning ALT relates to the final box in Figure 7.10. In this portion of the textbook, instances were identified in which students are "expected to identify patterns in, mainly numeric, contexts as representing linear, exponential, quadratic, or other relationships" (Olson, 2010, p. 286).

Figure 7.11 summarizes the working visual model for the textbook-spanning ALT in the *CMP2* grade 8 textbook, as well as the functional relationships ALT and the associated underlying ALTs. The grey arrows across the top of the model relate to the textbook-spanning ALT. The first three arrows in the textbook-spanning ALT define the functional relationship ALT and are thusly predominantly related to linear, exponential, and quadratic

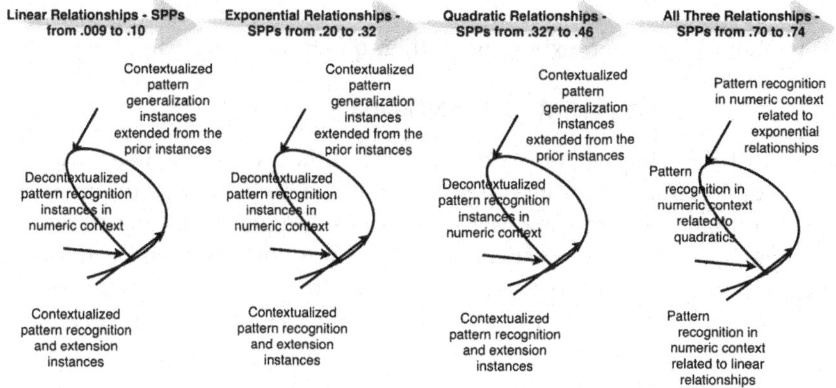

FIGURE 7.11. Visual model for ALT for *CMP2,* Grade 8. *Note:* A visual model for an identified ALT spanning *CMP2,* Grade 8. The four large grey arrows across the top of this model represent the ALT that spans the textbook. Included in this model are the underlying ALTs that were identified within each of the four sections of the main textbook-spanning ALT. The three arrows in the underlying ALTs represent three overall content trends within these ALTs and should be read as proceeding from the bottom description counter clockwise among the three descriptions. The descriptions were added to highlight the nature of the content. The related curriculum SPP spans were included to indicate in which part of the textbook one should expect to find the concepts presented in this model.

relationships, respectively. The fourth arrow relates to the portion of the textbook-spanning ALT in which instances were found that relate to syntheses of the three function relationships. The underlying ALTs represented circularly below the functional relationships ALT (the first three grey arrows) are to be read from the bottom and counterclockwise and indicate the way in which instances were generally found to follow a pattern. The first instances of these underlying ALTs involved contextualized pattern recognition and extension constructs, then instances included contextualized pattern generalization constructs, and finally instances were comprised of decontextualized pattern recognition problems in predominantly numeric contexts.

The circularly represented underlying ALT below the fourth grey arrow (to be read from the bottom and counterclockwise) represents the way in which the instances in the final stage of the overarching ALT relate first to liner representations, then exponential representations, and lastly quadratic representations. This underlying ALT related to the fourth grey arrow is not necessarily a well-defined progression: Often, instances involved all three function relationships. However, in general, the progression represented is aligned to the overall content thrust of the textbook-spanning ALT: namely, linear, then exponential, then quadratic relationships.

IMPLICATIONS AND DISCUSSION

The goal of this chapter was to demonstrate how, by conducting mathematical content analyses using the framework and methodology presented here, meaningful and informative differences among textbooks were identified, key aspects of mathematical landscapes were mapped, and working visual models of ALTs were developed. In this final section, implications and discussions related to the ALT-based mathematical content analysis are presented. Four potential areas for further research are explored: investigations of the implemented curriculum (including issues regarding supplementation), explication of students' development of mathematical knowledge in relation to learning trajectory constructs, articulation of the extent of alignment between textbooks and standards and assessments, and the implications for both preservice and in-service teacher education programs.

Implications for Implemented Curriculum

In commentary on what is known regarding the relationship between teachers and the curriculum materials they use, Remillard (2009) argued, "Regardless of how these decisions were made, the fact remains that each curriculum program has a number of features (some of which may seem trivial to the researcher or developer) that might figure significantly in how a teacher interacts with it" (p. 90). Differences in features between a textbook a teacher used the previous year, and the one she or he has been asked to use this year can potentially create enough discomfort mathematically for the teacher that the teacher decides to supplement the new textbook.

The possibility exists (as was seen in the grade 6 books of the *Glencoe* and *MathThematics* series) that content supplemented early in one book might simply be in a different location within a differently sequenced ALT in the new textbook. As such, there appears to be great potential for the use of such a framework to inform teachers' own comparison of textbook content sequencing as they plan their implementation of a new curriculum, as well as a research investigation of curriculum implementation and supplementation.

Recall the ways in which *Glencoe* and *MathThematics* grade 6 textbooks differed with respect to when pattern concepts were presented in the textbook (see Figures 7.4 and 7.5, and the related analyses). What information does a teacher need to know about how her textbook differs from others, particularly if students are transient within a district with no textbook adoption policy? If a student transfers from a *Glencoe,* grade 6 class to a *MathThemat-*

ics, grade 6 class about one fourth of the way through the textbook, then that student might not be exposed to pattern concepts and/or sequence concepts in grade 6. Teachers, in coordination with mathematics educators or mathematicians, could conduct parts or all of this framework for ALT-based content analyses related to various textbook series simply to foster awareness of the potential for these issues to arise and to facilitate discussion of possible strategies for dealing with such situations before they arise.

Implications for Learning Trajectory-Based Studies of Students' Knowledge Development

Researchers utilize learning trajectory constructs to investigate the ways in which students build mathematical knowledge (Confrey et al., 2009) and develop "empirically based models of children's thinking and learning" (Clements, 2007, p. 45) to inform curriculum research and design. Delineation of textbook ALTs could provide vital information regarding students' mathematical experiences. Furthermore, the negotiations between working visual models of identified ALTs and empirically based models could meaningfully support teachers' decisions to supplement textbook materials and/or sequence the textbook material in ways other than those suggested by the chapter order.

In constructing an empirical model of the ways middle school students develop mathematical knowledge associated with algebraic thinking and function concepts, one would need to study middle school students. Part of a student's mathematical experience is the negotiation of the mathematical content presented in the textbook from which she or he studies. In other words, to understand middle school students' mathematical knowledge development, it seems necessary to understand the span of mathematical content (e.g., as defined by the textbook authors) that the student has had to integrate and/or reject as he or she develops viable mental constructs regarding that mathematics content. If we assume that a student's mathematical experience includes the study of mathematics from existing curricula, then understanding the nature of these particular mathematical experiences vis-à-vis identified ALTs associated with students' experiences can inform the already meaningful work being done on elucidating the empirical learning trajectories associated with the ways in which students develop mathematical knowledge.

Implications for Alignment Studies

In the summary report of a conference related to the recent release of the Common Core State Standards for Mathematics (CCSS-M) (National Governors Association Center for Best Practices and Council of Chief State School Officers, 2010), Confrey and Krupa (2010) identified several recommendations to guide both the process of state adoption of the CCSS-M, and the development of future curricula consistent with the CCSS. In particular, Recommendation 9 of their report states, "Revise existing curriculum materials/resources to reflect the changes specified in the CCSS" (p. 14). One of the action items related to Recommendation 9 is, "Perform a content analysis of existing curricula, identifying areas in deed of additional development and/or modification to strengthen alignment and coherence" (p. 14).

With respect to that action item, ALT-based mathematical content analyses could serve as a framework for conducting meaningful and informative analyses. Regarding the framework presented in this chapter, the identification of content around which to focus a study utilizing this framework could be guided by the learning trajectories in the CCSS-M document. Analyses could be conducted on various curricula to determine alignment, not only to ensure that the appropriate content is present in the grade level textbook for each standard, but identifying where that content occurs within the textbook as well—the *mathematical landscape* of the CCSS-M in textbooks.

Implications for Teacher Education

There are currently efforts in the field to integrate curriculum analysis tasks into the experience of students enrolled in mathematics education courses and to integrate middle-grades textbook content into content courses taken by preservice and in-service teacher education students. One goal of these efforts has been identified as developing mathematics teachers' *curricular reasoning* (Breyfogle, McDuffie & Wohlhuter, 2010; Roth McDuffie & Mather, 2009). In particular, Breyfogle et al. (2010) identify the analysis of textbooks as an activity that specifically develops curricular reasoning.

Although teacher education programs may use similar textbook analysis projects as course requirements, it is unlikely that a consistent framework for analysis within these projects is employed. However, by using the ALT construct as an orienting framework for textbook analysis projects, the potential exists for the students in teacher education programs to have a consistent experience related to investigating the span of mathematical content (including meaningfully and theoretically grounded discussions of vertical alignment, prior knowledge, and so forth).

CONCLUSION

In this chapter, I presented a framework for mathematical content analyses of textbooks utilizing the articulated learning trajectory construct. There currently is little principle-based analysis of the sequencing of mathematical topics and concepts in textbooks and of the implications different sequencing has for mathematical alignment among textbooks, mathematical alignment between textbooks and standards documents, and perhaps most important, mathematical alignment between certain ALTs and high-stakes assessments of learning outcomes.

The use of a well-defined, theoretically based framework for conducting mathematical content analyses could lead to more meaningful and useable information that can be systematically distributed and used in the field. Furthermore, I believe that this information will have a positive impact on various facets of the mathematics education (and mathematics) community, including the areas of curriculum research, theoretical development of student learning, alignment with standards documents, and teacher education.

AUTHOR NOTE

This research was supported with funding for Doctoral Fellowship support through the Center for the Study of Mathematics Curriculum, a Center for Learning and Teaching funded by the National Science Foundation under Grant 0958058. Views expressed here are those of the author only, and do not necessarily represent the views of the National Science Foundation. The author wishes to thank his doctoral advisor, Dr. Douglas A. Grouws, for his guidance, support, and critical feedback throughout the study. The author also wishes to thank the editors of this volume, Drs. Alan Maloney, Jere Confrey, and Kenny Nguyen for their patient, thoughtful, and constructive feedback throughout the submission and revision process.

NOTES

1. Differences can be delineated among the terms textbooks, textbook curriculum, curriculum materials, curricular programs, and so forth. Unless otherwise noted, these terms are used interchangeably to refer to student edition textbooks.
2. Student edition textbooks were the only focus of the analysis.
3. *CMP2* is published in two formats: individual modules and modules bound in grade-level textbooks. The grade-level textbooks were analyzed in this study.
4. For distinctions between the terms *algebra* and *algebraic thinking*, see the synthesis provided in Olson (2010) (also see Bass, 2005; Chazan, 2008; Greenes & Rubenstein, 2008; Howe, 2005; Kilpatrick & Izsak, 2008; National Mathematics Advisory Panel, 2008). The categorization provided by Smith (2003) was the primary distinction utilized for this study: "Because the use of symbol systems has become prevalent, the word *algebra*, has become...synonymous

with the study or use of such systems. *Algebraic thinking*...has been used...to indicate the kinds of generalizing that precede or accompany the use of algebra" (p. 138).

5. A relevant page is any page from "the opening of the first chapter through the last chapter test" (Flanders, 1994, p. 68). Interestingly, this classification omits supplementary exercises at the end of texts, glossaries, appendices, answer pages, and indices.
6. For in-depth discussion and analysis of reliability work associated with the study reported in this chapter, see the Appendix of Olson (2010).
7. The *CMP2* textbooks used in this study posed a unique challenge in determining SPPs, as they are individually bound according to grade level (e.g., *CMP2* Grade 6, *CMP2* Grade 7, and *CMP2* Grade 8). However, the grade-level textbooks are composed of separate unit textbooks, each with individual page enumerations, glossaries, and other material extraneous to the focus of this study. To compute the SPP coefficient for a particular grade-level textbook page, it was assumed that the intended order of content was defined by the way in which the books were bound. This assumption was made because the unit textbooks came, directly from representatives of the publisher, bound in a particular order at each grade level. Thus, the SPPs associated with *CMP2* represent the percentage of the way through the textbook that an instance occurs without regard to extraneous materials (as defined by Flanders).
8. For an in-depth description of the ALTs identified within and across the four textbook series, refer to the extensive narratives developed in Olson (2010). It is from these narratives and the graphs presented in this chapter that the working visual models were developed.

REFERENCES

Adams, L. M., Tung, K. K., Warfield, V. M., Knaub, K., Mudavanhu, B., & Yong, D. (2000). *Middle school mathematics comparisons for Singapore mathematics, connected mathematics program, and mathematics in context (including comparisons with the NCTM principles and standards 2000).* Report to the National Science Foundation. Unpublished manuscript.

American Association for the Advancement of Science. (2000). *Middle grades mathematics textbooks: A benchmarks-based evaluation.* Retrieved from http://www.project2061.org/publications/textbook/mgmth/report/default.htm

Bailey, R., Day, R., Frey, P., Howard, A. C., Hutchens, D. T., McClain, K., . . . Willard, T. (2006). *Mathematics: Applications and concepts, courses 1 through 3.* New York, NY: Glencoe/McGraw-Hill.

Baroody, A. J., Cibulskiks, M., Lai, M., & Li, X. (2004). Comments on the use of learning trajectories in curriculum development and research. *Mathematical Thinking and Learning, 6*(2), 227–260.

Bass, H. (2005, September). *Review of the 4th and 8th grade algebra and functions items on NAEP.* Retrieved from http://www.brookings.edu/gs/brown/algebraicreasoning/Bass_Presentation.pdf

Battista, M. T. (2004). Applying cognition-based assessment to elementary school students' development of understanding of area and volume measurement. *Mathematical Thinking and Learning, 6*(2), 185–204.

Billstein, R., & Williamson, J. (2008). *MathThematics: Books 1 through 3.* Evanston, IL: McDougal Littell.

Breyfogle, M. L., McDuffie, A. R., & Wohlhuter, K. A. (2010). Developing curricular reasoning for grades pre-k-12 mathematics instruction. In B. J. Reys & R. E. Reys (Eds.), *Mathematics curriculum: Issues, trends, and future directions* (pp. 307–320). Reston, VA: National Council of Teachers of Mathematics.

Brown, A. L. (1992). Design experiments: Theoretical and methodological challenges in creating complex interventions in classroom settings. *Journal of the Learning Sciences, 2*(2), 141–178.

Chávez, O. (2003). *From the textbook to the enacted curriculum: Textbook use in the middle school mathematics classroom.* Unpublished doctoral dissertation, University of Missouri, Columbus, MO.

Chazan, D. (2008). The shifting landscape of school algebra in the United States. In C. E. Greenes & R. Rubenstein (Eds.), *Algebra and algebraic thinking in school mathematics: Seventieth yearbook of the NCTM* (pp. 19–33). Reston, VA: National Council of Teachers of Mathematics.

Clements, D. H. (2002). Linking research and curriculum development. In L. D. English (Ed.), *Handbook of international research in mathematics education* (pp. 599–630). Mahwah, NJ: Lawrence Erlbaum Associates.

Clements, D. H. (2007). Curriculum research: Toward a framework for "research-based curricula." *Journal for Research in Mathematics Education, 38*(1), 35–70.

Clements, D. H., & Sarama, J. (2004). Learning trajectories in mathematics education. *Mathematical Thinking and Learning, 6*(2), 81–89.

Clements, D. H., Wilson, D. C., & Sarama, J. (2004). Young children's composition of geometric figures: A learning trajectory. *Mathematical Thinking and Learning, 6*(2), 163–184.

Clopton, P., McKeown, E., McKeown, M., & Clopton, J. (1999). *Mathematically correct mathematics program reviews for grades 2, 5, and 7.* Retrieved from http://mathematicallycorrect.com/books.htm

Cobb, P., Confrey, J., diSessa, A., Lehrer, R., & Schauble, L. (2003). Design experiments in educational research. *Educational Researcher, 32*(1), 9–13.

Collins, J. V. (1911). *Practical algebra: Second course.* New York, NY: American Book Company.

Confrey, J. (2011). Better measurement of higher-cognitive processes through learning trajectories and diagnostic assessment in mathematics: The challenges in adolescence. In V. F. Reyna, S. B. Chapman, M. R. Dougherty, & J. Confrey (Eds.), *The adolescent brain: Learning, reasoning, and decision making.* Washington, DC: American Psychological Association.

Confrey, J., & Krupa, E. (2010). *Curriculum design, development, and implementation in an era of common core state standards: Summary report of a conference.* Retrieved from http://mathcurriculumcenter.org/conferences/ccss/SummaryReport-CCSS

Confrey, J., Maloney, A., Nguyen, K., Mojica, G., & Myers, M. (2009). Equipartitioning/splitting as a foundation of rational number reasoning using learning trajectories. In M. Tzekaki, M. Kaldrimidou, & H. Sakonidis (Eds.), *Proceedings of the 33rd conference of the International Group for the Psychology of Mathematics Education* (Vol. 2, pp. 345–352). Thessaloniki, Greece: International Group for the Psychology of Mathematics Education.

Drake, C. (2010). Understanding teachers' strategies for supplementing textbooks. In B. J. Reys & R. E. Reys (Eds.), *Mathematics curriculum: Issues, trends, and future directions* (pp. 277–287). Reston, VA: National Council of Teachers of Mathematics.

Fey, J. T. (1990). Quantity. In L. A. Steen (Ed.), *On the shoulders of giants: New approaches to numeracy* (pp. 61–94). Washington, DC: National Academy Press.

Flanders, J. (1994). Student opportunities in grade 8 mathematics: Textbook coverage of the SIMS test. In I. Westbury, C. A. Ethington, L. A. Sosniak & D. P. Baker (Eds.), *In search of more effective mathematics education: Examining data from the IEA second international mathematics study* (pp. 61–93). Norwood, NJ: Ablex.

Friedlander, A., Hershkowitz, R., & Arcavi, A. (1989). Incipient "algebraic" thinking in pre-algebra students. In *Proceedings of the 13th conference of the International Group for the Psychology of Mathematics Education* (Vol. 1, pp. 283–290). Paris: International Group for the Psychology of Mathematics Education.

Ginsberg, A., Leinwand, S., Anstrom, T., & Pollock, E. (2005). *What the United States can learn from Singapore's world-class mathematics system (and what Singapore can learn from the United States): An exploratory study.* Washington, DC: American Institutes for Research.

Gravemeijer, K. (2004). Local instruction theories as means of support for teachers in reform mathematics education. *Mathematical Thinking and Learning, 6*(2), 105–128.

Greenes, C. E., & Rubenstein, R. (Eds.). (2008). *Algebra and algebraic thinking in school mathematics: Seventieth yearbook of the NCTM.* Reston, VA: National Council of the Teachers of Mathematics.

Grouws, D. A., & Cebulla, K. J. (2000). Elementary and middle school mathematics at the crossroads. In T. L. Good (Ed.), *American education, yesterday, today, and tomorrow* (Vol. 2, pp. 209–255). Chicago, IL: University of Chicago Press.

Grouws, D. A., Smith, M. S., & Sztajn, P. (2004). The preparation and teaching practices of United States mathematics teachers: Grades 4 and 8. In P. Kloosterman & F. K. Lester (Eds.), *Results and interpretations of the 1990 through 2000 mathematics assessments of the National Assessment of Educational Progress* (pp. 221–267). Reston, VA: National Council of Teachers of Mathematics.

Hake, S. (2007). *Saxon math: Courses 1 through 3.* Austin, TX: Harcourt Achieve.

Howe, R. (2005, September). *Comments on NAEP algebra problems.* Retrieved from http://www.brookings.edu/gs/brown/algebraicreasoning/Howe_Presentation.pdf

Huntley, M. A., & Chval, K. (2010). Teachers' perspectives on fidelity of implementation to textbooks. In B. J. Reys & R. E. Reys (Eds.), *Mathematics curriculum: Issues, trends, and future directions* (pp. 289–304). Reston, VA: National Council of Teachers of Mathematics.

Kilpatrick, J., & Izsak, A. (2008). A history of algebra in the school curriculum. In C. E. Greenes & R. Rubenstein (Eds.), *Algebra and algebraic thinking in school mathematics: Seventieth yearbook of the NCTM* (pp. 3–18). Reston, VA: National Council of Teachers of Mathematics.

Lannin, J. K. (2003). Developing algebraic reasoning through generalization. *Mathematics Teaching in the Middle School, 8*(7), 342–348.

Lannin, J. K. (2005). Generalization and justification: The challenge of introducing algebraic reasoning through patterning activities. *Mathematical Thinking and Learning, 7*(3), 231–258.

Lannin, J. K., Barker, D. D., & Townsend, B. E. (2006a). Algebraic generalisation strategies: Factors influencing student strategy selection. *Mathematics Education Research Journal, 18*(3), 3–28.

Lannin, J. K., Barker, D. D., & Townsend, B. E. (2006b). Recursive and explicit rules: How can we build student algebraic reasoning? *The Journal of Mathematical Behavior, 25*(4), 299–317.

Lannin, J. K., Townsend, B. E., Armer, N., Green, S., & Schneider, J. (2008). Developing meaning for algebraic symbols: Possibilities & pitfalls. *Mathematics Teaching in the Middle School, 13*(8), 478–483.

Lappan, G., Fey, J. T., Fitzgerald, W. M., Friel, S. N., & Phillips, E. D. (2009). *Connected mathematics 2: Grades 6 through 8.* Boston, MA: Pearson Prentice Hall.

Lehrer, R., Kim, M.-J., & Schauble, L. (2007). Supporting the development of conceptions of statistics by engaging students in measuring and modeling variability. *International Journal of Computers for Mathematical Learning, 12*(3), 195–216.

Lesh, R., & Yoon, C. (2004). Evolving communities of mind-In which development involves several interacting and simultaneously developing strands. *Mathematical Thinking and Learning, 6*(2), 205–226.

MacGregor, M., & Stacey, K. (1992). A comparison of pattern-based and equation-solving approaches to algebra. In B. Southwell, K. Owens & B. Penny (Eds.), *Proceedings of the 15th annual conference, Mathematics Education Research Group of Australasia (MERGA)* (pp. 362–370). University of Western Sydney, Penrith, NSW: MERGA.

MacGregor, M., & Stacey, K. (1993). Seeing a pattern and writing a rule. In I. Hirabayshi, N. Nohda, K. Shigematsu, & F. Lin (Eds.), *Proceedings of the 17th conference of the International Group for the Psychology of Mathematics Education* (Vol. 1, pp. 181–188). University of Tsukuba, Japan: International Group for the Psychology of Mathematics Education.

MacGregor, M., & Stacey, K. (1997). Students' understanding of algebraic notation. *Educational Studies in Mathematics, 33*(1), 1–19.

McNaught, M., Tarr, J. E., & Grouws, D. A. (2008, March). *Assessing curriculum implementation: insights from the comparing options in secondary mathematics: Investigating curriculum (COSMIC) project.* Paper presented at the annual meeting of the American Educational Research Association, New York, NY.

Mullis, I. V. S., Martin, M. O., Gonzalez, E. J., Gregory, K. D., Garden, R. A., O'Connor, K. M., . . . Smith, T. A. (2000). *TIMSS 1999 international mathematics report: Findings from IEA's repeat of the Third International Mathematics and Science Study at the eighth grade.* Chestnut Hill, MA: International Study Center Lynch School of Education Boston College.

National Council of Teachers of Mathematics. (1989). *Curriculum and evaluation standards for school mathematics.* Reston, VA: Author.

National Council of Teachers of Mathematics. (2000). *Principles and standards for school mathematics.* Reston, VA: Author.

National Governors Association Center for Best Practices and Council of Chief State School Officers. (2010). *Common core state standards for mathematics.* Washington, DC: Authors.

National Mathematics Advisory Panel. (2008). *Foundations for success: The final report of the National Mathematics Advisory Panel.* Washington, DC: U.S. Department of Education.

National Research Council. (2001). *Adding it up: Helping children learn mathematics* (J. Kilpatrick, J. Swafford, & B. Findell, Eds.). Washington, DC: National Academy Press.

National Research Council. (2004). *On evaluating curricular effectiveness: Judging the quality of K-12 mathematics evaluations.* Washington, DC: National Academy Press.

Olson, T. A. (2010). *Articulated learning trajectories related to the development of algebraic thinking that follow from patterning concepts in middle grades mathematics.* Unpublished doctoral dissertation, University of Missouri, Columbia, MO. Retrieved from https://mospace.umsystem.edu/xmlui/handle/10355/8318

Olson, T. A., Regis, T. P., & Papick, I. J. (2010). Pattern problems in middle grade mathematics curricula. *Investigations in Mathematics Learning, 2*(3), 1–23.

Papick, I. J., Olson, T. A., & Regis, T. P. (2010). Analyzing numerical and geometric pattern problems in middle grade mathematics curricula. *Investigations in Mathematics Learning, 2*(3), 24–42.

Radford, L. (1999). The rhetoric of generalization. In O. Zaslavsky (Ed.), *Proceedings of the 23rd conference of the International Group for the Psychology of Mathematics Education* (Vol. 4, pp. 89–96). Haifa, Israel: International Group for the Psychology of Mathematics Education.

Radford, L. (2000). Signs and meanings in students' emergent algebraic thinking: A semiotic analysis. *Educational Studies in Mathematics, 42*(3), 237–268.

Radford, L., Bardini, C., & Sabena, C. (2007). Perceiving the general: The multidimension of students' algebraic activity. *Journal for Research in Mathematics Education, 38*(5), 507–530.

Remillard, J. T. (2009). Part II commentary: Considering what we know about the relationship between teachers and curriculum materials. In J. T. Remillard, B. A. Herbel-Eisenmann & G. M. Lloyd (Eds.), *Mathematics teachers at work: Connecting curriculum materials and classroom instruction* (pp. 85–92). New York, NY: Routledge.

Roth McDuffie, A., & Mather, M. (2009). Middle school mathematics teachers' use of curricular reasoning in a collaborative professional development project. In J. T. Remillard, B. A. Herbel-Eisenmann & G. M. Lloyd (Eds.), *Mathematics teachers at work: Connecting curriculum materials and classroom instruction* (pp. 302–320). New York, NY: Routledge.

Sabena, C., Radford, L., & Bardini, C. (2005). Synchronizing gestures, words and actions in pattern generalizations. In H. L. Chick & J. L. Vincent (Eds.), *Proceedings of the 29th conference of the International Group for the Psychology of Mathematics Education* (Vol. 4, pp. 129–136). Melbourne, Australia: International Group for the Psychology of Mathematics Education.

Sasman, M. C., Olivier, A., & Linchevski, L. (1999). Factors influencing student's generalisation thinking processes. In O. Zaslavsky (Ed.), *Proceedings of the 23rd conference of the International Group for the Psychology of Mathematics Education* (Vol. 4, pp. 161–168). Haifa, Israel: PME.

Saul, M. (2008). Algebra: The mathematics and the pedagogy. In C. E. Greenes & R. Rubenstein (Eds.), *Algebra and algebraic thinking in school mathematics: Seventieth yearbook of the NCTM* (pp. 63–79). Reston, VA: National Council of Teachers of Mathematics.

Senechal, M. (1990). Shape. In L. A. Steen (Ed.), *On the shoulders of giants: New approaches to numeracy* (pp. 139–181). Washington, DC: National Academy Press.

Simon, M. A. (1995). Reconstructing mathematics pedagogy from a constructivist perspective. *Journal for Research in Mathematics Education, 26*(2), 114–145.

Simon, M. A., & Tzur, R. (2004). Explicating the role of mathematical tasks in conceptual learning: An elaboration of the hypothetical learning trajectory. *Mathematical Thinking and Learning, 6*(2), 91–104.

Smith, C. (1924). *A treatise on algebra.* London, UK: MacMillian.

Smith, E. (2003). Stasis and change: Integrating patterns, functions, and algebra throughout the K–12 curriculum. In J. Kilpatrick, W. G. Martin & D. Schifter (Eds.), *A research companion to principles and standards for school mathematics* (pp. 136–150). Reston, VA: National Council of Teachers of Mathematics.

Steen, L. A. (1990). Pattern. In L. A. Steen (Ed.), *On the shoulders of giants: New approaches to numeracy* (pp. 1–10). Washington, DC: National Academy Press.

Steffe, L. P. (2004). On the construction of learning trajectories of children: The case of commensurate fractions. *Mathematical Thinking and Learning, 6*(2), 129–162.

Stein, M. K., Remillard, J., & Smith, M. S. (2007). How curriculum influences student learning. In F. K. Lester, Jr. (Ed.), *Second handbook of research on mathematics teaching and learning: A project of the National Council of Teachers of Mathematics* (pp. 319–369). Charlotte, NC: Information Age Publishing.

Swafford, J. O., & Langrall, C. W. (2000). Grade 6 students' preinstructional use of equations to describe and represent problem situations. *Journal for Research in Mathematics Education, 31*(1), 89–112.

Taplin, M. L., & Robertson, M. E. (1995). Spatial patterning: A pilot study of pattern formation and generalisation. In L. Meira & D. Carraher (Eds.), *Proceedings of the 19th conference of the International Group for the Psychology of Mathematics Education* (Vol. 3, pp. 42–49). Recife, Brazil: PME.

Tarr, J. E., Chávez, O., Reys, R. E., & Reys, B. J. (2006). From the written to the enacted curricula: The intermediary role of middle school mathematics teachers in shaping students' opportunity to learn. *School Science and Mathematics, 106*(4), 191–201.

Tyson, H., & Woodward, A. (1989). Why students aren't learning very much from textbooks. *Educational Leadership, 47*(3), 14–17.

U.S. Department of Education's Mathematics and Science Expert Panel. (1999). *Exemplary and promising mathematics programs.* Washington, DC: U.S. Department of Education.

Wells, W. (1908). *Algebra for secondary schools.* Boston, MA: D.C. Heath.

Woodward, A., & Elliot, D. L. (1990). Textbook use and teacher professionalism. In D. L. Elliot & A. Woodward (Eds.), *Textbooks and schooling in the United States (89th yearbook of the National Society for the Student of Education, part 1)* (pp. 178–193). Chicago, IL: University of Chicago Press.

CHAPTER 8

LEARNING TRAJECTORIES AND PROFESSIONAL DEVELOPMENT

P. Holt Wilson

ABSTRACT

This chapter reports on how a group of teachers came to understand and apply the concept of a learning trajectory in the context of early rational number reasoning. It identifies transitions in knowledge experienced by participating teachers and changes in their practice that resulted from their understandings. These included increased ability to create models of student thinking, deepening of their own content knowledge, informing their instructional planning and assessment, and promoting more targeted discourse.

LEARNING TRAJECTORIES AND PROFESSIONAL DEVELOPMENT

As learning trajectories continue to increase in importance in education, it is imperative to consider their utility for teachers, the implications of their appropriation by teachers, and the types of supports teachers need to use them. Learning trajectories have been shown to inform curriculum devel-

Learning Over Time: Learning Trajectories in Mathematics Education, pages 227–142.

opment (Clements, Wilson, & Sarama, 2004; see also Chapters 1 and 2, this volume), assessment design (Battista, 2004), and the articulation of standards (see Chapter 5, this volume), all with the goal of improved student achievement. It is essential to keep in mind that classroom teachers are the key links between systemic innovation and improved student learning.

Cohen and Hill (2001) remind us that targeted professional development (PD) for teachers is essential for effective large-scale reform initiatives. In studying the adoption and implementation of the California Mathematics Framework in the 1990s, they found that absent teachers' opportunities to learn about the standards, only modest gains in student achievement were observed. General PD opportunities were associated with increased student mathematics achievement, but the greatest gains occurred with students whose teachers had PD opportunities focused on engaging them in elaborations of the standards and in the various ways in which students approach mathematical concepts.

This chapter reports research on PD using a learning trajectory to assist practicing teachers in understanding changes in state standards. As a part of the ongoing work of the Diagnostic e-Learning Trajectories Approach (DELTA) project at North Carolina State University, I designed PD to provide opportunities for elementary mathematics teachers to understand the K–2 Number & Operations and Geometry strands of the NC Essential Standards through the framework of an *equipartitioning learning trajectory* (ELT) (see Chapter 3, this volume) and to relate this understanding to their classroom instruction.

This chapter first sets out an overview of the PD and the design of the research component of the study. Findings from this research are discussed with regard to two major themes: first, the potential for learning trajectories to support transitions in teachers' understanding of their own and of students' knowledge of content, and second, the ways that learning trajectories may influence teachers' instructional practices. The chapter concludes with a summary of effective learning activities for PD on learning trajectories and identification of several challenges associated with supporting teachers' learning and use of learning trajectories.

EARLY RATIONAL NUMBER REASONING: PROFESSIONAL DEVELOPMENT DESIGN

In the early stages of the DELTA project's work in constructing learning trajectories as a basis of a diagnostic assessment system, it became clear that the system's design needed to be informed by a "use model" (Confrey & Maloney, 2012)—the ways in which teachers would use diagnostic information in their classrooms. As a related dissertation study, I used a PD context to investigate ways in which teachers came to understand learning trajectories and how they used learning trajectories to inform their planning, instruction, and assessment. One of the aims of the dissertation study was to develop insight into the types of PD that would be needed to support teachers' use of learning trajectories to inform their instruction and to provide diagnostic information for instructional guidance.

The PD, entitled Early Rational Number Reasoning for K–2 Teachers (ERNR PD), had three main goals:

1. To assist teachers in understanding and using a learning trajectory for an early rational number construct called equipartitioning (see Chapter 3, this volume)
2. To equip teachers with strategies to recognize and use students' thinking in their practice
3. To support their students' learning of early rational number concepts

Overview of the Professional Development

The PD consisted of 20 hours comprised of 10 hours of face-to-face instruction and 10 hours of classroom-based activities. The face-to-face instruction focused on the ELT and on instructional practices that would enhance teachers' understandings of students' mathematical thinking, including clinical interviewing, instructional task analysis and adaptation, formative assessment using artifacts of students' work, and principles of classroom discourse. The classroom activities permitted teachers to engage in these strategies with their students, their colleagues, and the curriculum. Table 8.1 provides an overview of the ERNR PD.

Session I focused on carefully listening to and observing students, with the goal of understanding students' ideas rather than judging them as right or wrong. In this session, teachers learned how to conduct and analyze clinical interviews to diagnose students' understandings. In the subsequent first classroom activity, teachers conducted and videotaped an interview with one of their students involving the task of fairly sharing a collection of 24 gold coins among different numbers of pirates. After reviewing their recordings, teachers published their analyses to a blog that was read and commented on by other ERNR participants.

In Session II, teachers were introduced to the idea of a learning trajectory, studied the portion of the ELT that pertains to sharing collections, and related those ideas to the new state standards. Recordings from the teachers' previous blog posts were used to illustrate various behaviors and verbalizations described in the ELT. Next, using selected activities from their curricular materials, the teachers learned to analyze instructional tasks, ascertain their mathematical goals, and adapt them to include equipartitioning. For the classroom activity that followed, teachers selected and analyzed

TABLE 8.1. Timeline and Activity Structure of the Study

Time period	Activity	Topic
Week 1	PD Session I	Clinical interview training
Week 2	Classroom Activity One	Clinical interview—Sharing a collection
Week 3	PD Session II	EPLT I: Sharing a collection
Week 4	Classroom Activity Two	Task adaptation
Week 5	PD Session III	EPLT II: Sharing a whole
Week 6	Classroom Activity Three	Clinical interview—Sharing a whole
Week 7	PD Session IV	EPLT III: Sharing multiple wholes
Weeks 8–13	Equipartitioning Instruction	Teaching lessons on fair sharing
Week 14	PD Session V	Summary and evaluation of the PD

a task from their curriculum that was amenable to sharing collections and adapted it. They posted their adaptations and rationales based on the ELT to the blog for peer review.

During Session III, teachers examined portions of the ELT pertaining to sharing a whole. They engaged in paper-folding tasks to better understand the complexities of equipartitioning a whole into various numbers of equal-sized parts as portrayed in the trajectory. They then discussed ideas of formative assessment and completed an activity in which they examined and analyzed student responses to tasks of sharing a whole. They undertook to describe both what they believed the students knew about equipartitioning and the types of feedback they would provide to the students. Their third classroom activity was to conduct a second interview with the same student from Session I, but this time, using a task of fairly sharing rectangular and circular wholes for various numbers of people. Teachers again videotaped, analyzed, and blogged about the student's understanding of equipartitioning, this time in relation to their knowledge of the ELT.

Session IV presented the remainder of the ELT, which pertains to sharing multiple wholes, to provide teachers with a sense of how the ideas of sharing collections and sharing single wholes supported subsequently more complex understandings of equipartitioning multiple wholes. Through an activity of sorting students' responses to various tasks of sharing multiple wholes, they learned the different strategies students use to complete these types of tasks, as well as related issues of referent units, improper fractions, and mixed numbers. After this session, a subset of the participants volunteered to teach lessons on fair sharing in their classrooms and to participate in an interview on instruction related to equipartitioning.

In Session V, the final session, teachers synthesized the three portions of the ELT they had studied and engaged in a series of cross-grade discussions on task selection and adaptation, student work analysis, and instruction. In a final clinical interview analysis activity, teachers viewed a clinical interview of a child fairly sharing a whole. During the viewing, the videotape was paused multiple times to allow teachers to summarize what they believed the child knew about fair sharing, the ways they anticipated the child would approach the next task, and to reflect on the types of instructional opportunities they might provide to support the child in progressing through the ELT.

ERNR PD RESEARCH DESIGN: ANTICIPATING A USE MODEL

The research in this study was designed to gain insights into the ways that, through the ERNR PD, teachers would come to understand and appropriate the ELT in their instruction, in order to assist the DELTA project in conceptualizing a use model for its developing diagnostic assessment system. Three questions guided the research:

1. In what ways and to what extent do teachers use the ELT to build models of students' thinking?
2. In what ways and to what extent do teachers use the ELT to inform their adaptations of curricular instructional tasks, their interactions with students during instruction, and their assessment of students' understanding?
3. What are the relationships among teachers' knowledge of equipartitioning, their uses of the ELT, and their students' learning?

Because a central idea of the learning trajectory construct is the development of student conceptual understanding, the study was framed using the idea of teachers as builders of theoretic models of students' mathematics understanding (Cobb & Steffe, 1983). Such theoretic models entail using a better-known domain of knowledge to understand the elements, structures, and systems of a lesser-known domain (Black, 1962). For instance, when teachers use their own understandings of the content along with the ways students formulate ideas, in order to make sense of students' approaches to particular mathematical problems, they are constructing a model of the students' thinking. To describe the ways that teachers' models evolved during the course of the PD, the study used Wilson, Lee, and Hollebrands's (2011) observations of teachers' model-building processes. These four processes include: (1) describing, in which teachers provide nonevaluative descriptions of students' actions and verbalizations; (2) comparing, in which teachers make explicit or implicit comparisons of students' actions to their own or to those of other students; (3) inferring, in which these descriptions and comparisons serve as evidence for inferences about students' cognition; and (4) restructuring, in which a teacher's own understandings of content and/or content in relation to students is accommodated. In the study, these processes were observed regarding teachers' planning, instruction, and assessment to investigate the ways in which a learning trajectory affects teachers' practices.

The research questions address the ways that teachers learn about learning trajectories as well as the mechanisms by which teachers acquire and appropriate their understandings of learning trajectories in their own practices; in short, the study investigated the processes by which teachers came to understand and use a learning trajectory and the relationships between those uses and students' learning. As noted by Confrey (2006), design research is an appropriate methodology when studying issues of process. Central to the method is the "'engineering' [of] particular forms of learning and systematically studying those forms of learning within the context defined by the means of supporting them" (Cobb, Confrey, diSessa, Lehrer, & Schauble, 2003, p. 9). Thus, the selection of design study as the research methodology for research on the ERNR PD implementation permitted the investigation of the ways that teachers came to understand and use learning trajectories in their practice.

In design studies, which have an iterative nature, researchers formulate initial conjectures about learning and then revisit them throughout the study's implementation to track changes in the conjectures with respect to what they are observing (Confrey & LaChance, 2000). Five conjectures, related to teachers' learning of the ELT and its relationships with students' learning, guided the implementation of this study. The conjectures asserted that knowledge of the ELT would (1) assist teachers in building models of students' thinking by highlighting facets of students' behaviors and language related to a particular conception; (2) inform teachers' adaptation of instructional tasks through adjusting the mathematical goals and judging the relative difficulty of tasks; (3) support teachers' formative assessment of students' work on instructional tasks by helping them identify evidence of students' thinking, predict what students may or may not know, and raise questions directing their next instructional steps; (4) influence teachers' instructional interactions with students through identifying student ideas that support the mathematical goals of equipartitioning, refining their own questions for students, and sequencing students' work during whole group discussions; and (5) be positively associated with their students' learning.

A school district in North Carolina had recently adopted *Investigations in Data, Number, and Space* (TERC, 2007) and was approached with a proposal to offer the ERNR PD to its teachers in order to relate *Investigations*, the forthcoming revised state mathematics standards informed by DELTA's work (see Chapter 5, this volume), and contemporary ELT research. Upon agreement with district leaders, I tailored the ERNR to support teachers in understanding the new standards by leading them through activities related to the ELT and aligning these with their newly adopted curriculum. A total of 33 K–2 teachers (11 kindergarten, 11 first grade, and 11 second grade) from two elementary schools in this rural district participated in the PD and the research. A subset of 10 second-grade teachers elected to participate in an observational study of classroom instruction.

Several primary sources of data were collected during the study, including an assessment of teachers' understanding of equipartitioning and related pedagogical issues administered at the beginning and at the conclusion of the PD (Mojica & Confrey, 2009); pilot diagnostic assessment items developed by the DELTA research team; video recordings of classroom instruction in equipartitioning and stimulated-recall interviews of the subset of teachers; and artifacts from the learning activities of the ERNR PD, including the clinical interview analyses (Classroom Activities 1 and 3), instructional task adaptations (Classroom Activity 2), the student work analysis activity in Session III, and the clinical interview analysis activity from Session V. Video and audio recordings of PD sessions and artifacts from other learning activities in Sessions I–IV were used as necessary to supplement interpretation of primary data.

Gain scores from the assessment of teachers' understanding of equipartitioning and the pilot diagnostic items were analyzed using Wilcoxon Sign Rank tests due to small sample sizes and non-normality of the sample distributions. Video, audio, classroom observation, and observation interview data were reviewed multiple times to identify critical moments (Powell, Francisco, & Maher, 2003), which were defined to be any instances related to the study's conjectures. Transcripts were created for each of these moments and were coded initially using the framework. From these categories, secondary coding using a constant comparison approach (Glaser, 1992) was conducted to reveal patterns within the data. Artifacts from learning activities from the PD were coded in a similar manner.

TRANSITIONS IN KNOWLEDGE

The findings from the research indicate that as teachers learned about the ELT during the course of the PD, their knowledge for teaching equipartitioning matured along two dimensions. First, their understandings of the ways in which students think about equipartitioning shifted from vague evaluations to more precise models of student thinking. Second, the teachers' own knowledge of equipartitioning developed through considering the different strategies and emergent relationships that children developed when engaging in the fair sharing tasks.

Building More Precise Models of Thinking

Prior to the ERNR PD, the teachers constructed models of students' thinking that were vague and yielded little information about students' cognition. That is, many teachers made general observations mostly related to the outcomes of tasks rather than to students' thinking that led to such outcomes. As they learned about the learning trajectory and about practices that elicited and used student thinking in instruction, the teachers' models began to express greater explanatory and anticipatory power in regard to linking students' behaviors and verbalizations with cognition. These shifts can be characterized using the model-building process categories used to frame the study: describing, comparing, inferring, and restructuring.

First, teachers' descriptions of student thinking at the beginning of the ERNR PD initially contained general or irrelevant observations. By the end of the PD, their descriptions tended to be more specific and focused on behaviors and verbalizations outlined in the ELT. Prior to experiences with the learning trajectory, teachers' descriptions were indistinct, such as noting that a student created two piles when sharing a collection of coins without commenting on whether the student had dealt the coins as singletons, dealt with composites, or counted and used known number facts. After experiences with the learning trajectory, however, they made clear distinctions among students' behaviors and verbalizations, such as noting that a student cut strips of paper from left to right when sharing a whole, rather than simply stating that the student created a certain number of parts.

Second, teachers made deeper comparisons when constructing their models of student thinking. Early on, they used referents that were exclusively from their past experiences with students and later incorporated references to theory detailed in the ELT. When modeling student thinking at the beginning of the PD, teachers compared students' work to recollections of former students or their own out-of-school experiences with children. After learning about the ELT, their comparisons incorporated references to the theory by which children's informal knowledge of equipartitioning developed over time, including the three equipartitioning criteria (see Chapter 3, this volume). Rather than make comparisons strictly based on age or grade level, the teachers became more oriented toward identifying which of the equipartitioning criteria were underdeveloped in students' solutions to problems.

Third, teachers' inferences transitioned from categorical judgments of the students' knowledge to more evidence-based and nuanced perceptions. At the beginning of the ERNR PD, many of the teachers made broad, unqualified evaluations of student understanding, and often simply stated that a student "understands" or "does not understand" equipartitioning. After work with the ELT, however, they commonly declared more specific claims about the productive conceptions held by students, and their focus shifted from solely observing outcomes to examining the processes that students used to arrive at an outcome. Moreover, they supported these claims with examples of specific behaviors and verbalizations they observed and which were described in the ELT, such as linking students' uses of particular equipartitioning strategies and the emerging relationships of compensation and composition to infer student cognition. Learning about the ELT stimulated alterations in initial dispositions toward students as deficient in their knowledge and reasoning into a generative one, in which teachers began to search for student ideas that would help the students grow in their understanding.

Finally, the teachers' experiences with the ELT affected the ways they restructured their own knowledge of equipartitioning, their pedagogical content knowledge related to equipartitioning, and portions of their pedagogical knowledge. Subsequent sections of this chapter elaborate on teachers' restructuring of each of these three domains. Thus, in building models of students' thinking, teachers used the ELT to enrich their explanations of students' performances based on the types of observations teachers make daily. The ELT served as a tool for teachers to coordinate students' behaviors and verbalizations with cognition. By providing a framework for conceptual development over time, the ELT enabled them to anticipate the ways students might engage with subsequent instruction.

Restructured Knowledge of Equipartitioning

Prior to beginning the ERNR PD, and then again upon its conclusion, teachers completed an assessment of equipartitioning knowledge developed by the DELTA team (Mojica & Confrey, 2009). Items on the instrument were quoted or adapted from the research literature and addressed two dimensions. Twelve items focused on specific mathematical content and were designed to identify ways that teachers themselves would solve equipartitioning tasks, such as predicting the number of equal-sized parts when a piece of paper is folded in half four times. Six of the items focused on broader, more flexible understandings of equipartitioning needed for teaching, such as generating multiple correct and incorrect solutions to the challenge of fairly sharing a circular birthday cake among six people (Mojica & Confrey, 2009).

An investigation of the gain scores for each of these two dimensions indicated modest increases resulting from the ERNR PD. For the items related to teachers' approaches to equipartitioning tasks, the median increase of 3 points (out of a possible 36) was statistically significant (**S** = 195.00, $p < 0.0001$). Likewise for the items related to equipartitioning and students' potential solutions, the median increase of 5 points (out of a possible 18) was also statistically significant (**S** = 243.00, $p < 0.0001$). The observations of participants' teaching of equipartitioning lessons supported and elaborated these findings.

Three distinct patterns related to teachers' understandings of equipartitioning emerged from these observations by the end of the study. Two represented misconceptions on the part of some of the teachers that persisted for the duration of the study, but one suggested ELT-based restructuring of teacher content knowledge. First, some of the teachers viewed equipartitioning as an outcome rather than an action of making equal-sized groups or parts. For instance, many teachers were unaware that many students "assembled" equal-sized parts to create a whole rather than making equal-sized parts from an initial whole, such as drawing adjoining equal-sized parts to form a "whole." Because the product of a student's work looked like a whole partitioned into equal-sized parts, some teachers deemed the solution correct. However, students' use of this approach does not maintain the relationship between the part and the original whole, which is the central idea of equipartitioning as a basis for rational number reasoning.

Second, a number of teachers were uncertain whether noncongruent parts could be equal-sized, particularly when creating four equal-sized parts of a rectangle using its diagonals. Three out of the ten teachers participating in the observational part of the study either failed to verify that this strategy produced fair shares or considered it incorrect when students presented the approach during instruction. This was surprising given that a significant exchange occurred during the ERNR PD concerning this specific example, complete with different justifications that students could offer to verify the equivalence of the parts.

Third, when teaching their lessons, some of the teachers identified and highlighted the emergent relationships described in the ELT. For example, many of the teachers facilitated classroom discussions related to compensation in the context of sharing among different numbers of people, noting that as the number of people sharing a whole increases, the size of the part decreases. Other teachers focused on the idea of composition when sharing a whole, demonstrating the equivalence of "one of four parts" and "two of eight parts." By using these relationships in instruction, the teachers highlighted ideas of equipartitioning that connect with issues of measurement (in the case of compensation) and ratio reasoning (in the case of composition and equivalence). Teachers' identification and use of emergent relationships suggests that structured PD experiences with a learning trajectory can lead to the restructuring of teachers' content knowledge.

INFLUENCES ON PRACTICE

The research findings also indicate that PD on the ELT resulted in teachers' restructuring their pedagogical content knowledge as well as their pedagogy. By the conclusion of the ERNR PD, teachers' understandings of equipartitioning in relation to both students' knowledge and teaching had improved, enriching their ability to observe and predict students' mathematical activity, to locate and move students along a continuum of conceptual development, and to discuss students' progress in equipartitioning with their colleagues in relation to *Investigations.* When the subset of teachers presented lessons on equipartitioning, the conceptual structure of the ELT supported some teachers in facilitating classroom discussions around ideas of equipartitioning using students' ideas.

Restructured Pedagogical Content Knowledge

Changes in the teachers' pedagogical content knowledge as a result of learning about the ELT manifested in three ways:

1. It sensitized them to behaviors indicative of different cognitive approaches and supported them in making predictions about students' behaviors and cognition.
2. It assisted them in locating students' thinking within the learning trajectory and influenced their next interactions with students.
3. It provided them with a language to discuss these behaviors even though it was largely insufficient in assisting them in explaining its interactions with their curriculum.

The data from the ERNR PD suggest that teachers became more sensitive to the subtleties and distinctions in students' behaviors and verbalizations. With the ELT providing a framework for coordinating their observations of students with the ways students might be thinking about equipartitioning, teachers were less likely to dismiss observations that they did not understand. This increased sensitivity coupled with the ELT allowed teachers to predict how students would engage in a task based on their previous work. For example, when participating in the clinical interview analysis activity (PD Session V), many teachers noticed that the child used repeated halving on a series of tasks. When asked to anticipate how the child would fairly share a circular cake among three people, these teachers suggested that the child might make four parts by first creating two equal-sized parts (halves) and then halving those halves, an approach that has in fact been witnessed in numerous clinical interviews.

By the conclusion of the ERNR PD, teachers were also able to situate a particular student's understanding within the range of conceptual understanding described by the ELT. Teachers used the ELT as a framework for describing ideas that a student was likely to already understand and which ideas were likely to develop next. By locating students' ideas within the trajectory, teachers were then able to make decisions about their next instructional steps based on this location and their knowledge of the ELT.

When a student experienced difficulty with equipartitioning, some teachers suggested changing task parameters such as the shape of the whole, while others suggested revisiting a previous level of understanding. Using the same example from the clinical interview analysis activity, one teacher suggested that the child was working toward understanding creating an odd number of equal-sized parts of a circular whole. When the child was unsuccessful, another teacher suggested that the student might be successful if the whole were a rectangle. Others indicated that the student might benefit from revisiting a previous level of understanding and sharing a collection among three people.

Throughout the PD, teachers developed and began to use specific vocabulary to describe and discuss students' behaviors, verbalizations, and cognition. For instance, many adopted and used mathematically precise terms like *exhausting the whole*, *radial* and *diametric* cuts, and *partitive* and *quotitive* division in their PD discussions. However, experiences with the ELT during the PD were mostly inadequate to support teachers in identifying and explaining interactions with the *Investigations* curriculum. Following an instructional unit pertaining to area measurement, one teacher provided 1-inch square tiles for her students to use in a lesson on fair-sharing rectangular wholes. She reported that she wanted her students to measure each of the parts created by equipartitioning to justify that they were of equal size. However, some students used the tiles to construct a rectangular whole, not maintaining the part–whole relationship that was the goal of the lesson. Though this teacher recognized the issue as it unfolded during instruction, teachers require more support to identify areas of curriculum that are amenable to—as well as those that may be problematic for—equipartitioning, to anticipate possible unintended consequences of integrating learning trajectories into existing curricula.

Restructured Pedagogical Knowledge

A final domain influenced by the ELT was the teachers' pedagogical knowledge. Specifically, the observation data suggest that three pedagogical practices related to facilitating coherent instruction were affected by the teachers' experiences with the ELT. Through stimulated-recall interviews on the implemented lessons, teachers indicated that their selections of specific student ideas to share during class discussion were influenced by the ELT. To a lesser degree, they reported that the way they sequenced those ideas, and the connections they made among them, were also related to the ELT.

Some of the teachers used the learning trajectory as a way to select examples to bring to the attention to the class. In the interviews, many teachers indicated that when circulating among students and choosing students to share work, they were looking for both correct and incorrect approaches or for a variety of strategies. Some reported that they were seeking solutions that would confront misconceptions related to equipartitioning, such as *n* splits makes *n* parts. Others were looking for different examples related to the three equipartitioning criteria. In one instance, a teacher selected a student solution to the task of fairly sharing a birthday cake among six friends that did not exhaust the whole. Through questioning the class about leftover cake, the student was able to progress to using the whole but with unequal-sized parts. Through subsequent careful questioning, the student ultimately created equal-sized parts. Teachers' uses of the ELT in selecting students' work varied. For most of the teachers, it served as a checklist of strategies to search for, identify, and exhibit in class. For a few, it was used as a resource for examining students' thinking and informing the selection of examples that would assist other students in their learning.

Though all teachers reported that they did not plan a specific order to the examples they asked students to share prior to facilitating the discussion, three of the teachers reported that as class discussion unfolded they used the ELT to order the sharing of ideas based on student presentations. For instance, after a pair of students presented their method for sharing a rectangular cake among four people, their teacher asked another pair to share their solution in which they created eight equal-sized parts and allocated each person two parts. When asked about her sequencing strategy in the observation interview, she reported that she recognized the opportunity to discuss equivalent fractions after the first student presentation. Though none of the teachers sequenced students' ideas from least to most sophisticated in relation to the ELT during instruction, data from the learning activities in the ERNR PD suggested that teachers were able to order solutions in this way and create questions to stimulate growth of understanding.

Finally, some of the teachers were able to connect students' ideas to support the class in deepening their understanding of equipartitioning. One teacher focused on confronting misconceptions by noting the three equipartitioning criteria across different student presentations of sharing a whole among various numbers of people. Another teacher overlaid transparencies of different students' sharing strategies for a rectangular whole among four and six people to develop the idea of compensation. Students used the same strategy to share among six as with four, as shown in Figure 8.1. By overlaying the transparencies that illustrated students' solutions, the teacher assisted the class in noticing that as more parts are created, the size of each part becomes smaller. Again, though most of the teachers did not make salient connections across different student presentations in their lessons, those who did so demonstrated that the ELT could provide a *student-centered* framework for instruction.

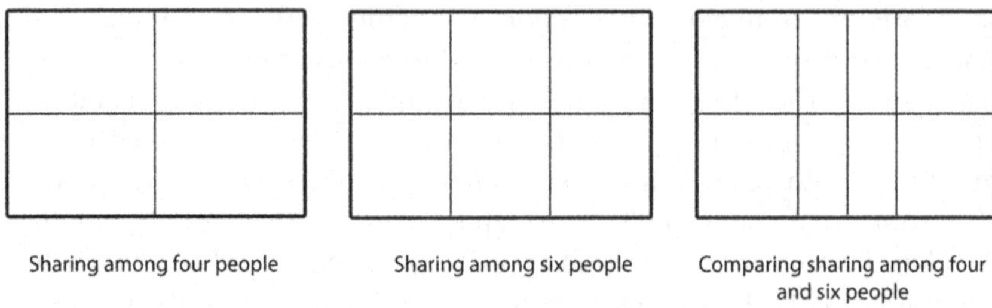

Sharing among four people
Sharing among six people
Comparing sharing among four and six people

FIGURE 8.1. One teacher's connection of student ideas in instruction.

DISCUSSION

Research on the ERNR PD provides a case for the utility of learning trajectories for teachers. Learning trajectories assist teachers in constructing more precise models of their students' thinking as well as in deepening teachers' own content knowledge. Additionally, learning trajectories support teachers in developing richer pedagogical content knowledge and can potentially be used as instructional guides or supplements to curriculum. While research in the ways teachers use learning trajectories is just beginning, other researchers are reporting similar findings in terms of transitions of knowledge (Furtak, 2009; Mojica, 2010).

As teacher educators design and implement PD related to learning trajectories, it is worth noting that several learning activities from the ERNR PD were notably effective in supporting teachers in understanding and using trajectories in practice based on the analysis of the data from the PD sessions. First, the use of video recordings to illustrate different behaviors and levels of reasoning sophistication was particularly effective in helping teachers identify and recall the specifics of the learning trajectory. Similarly, the process of analyzing student work was productive for teachers and led many of them to begin to move away from cursory evaluations of student thinking that were typical at the beginning of the PD to viewing students' conceptual development along a continuum of increasing adequacy and sophistication. Finally, viewing longer recordings of children engaged in tasks (e.g., a clinical interview) provided teachers with opportunities to construct and accommodate a model of thinking, to be surprised by the resourcefulness of children, and to reflect on how they might support children in continuing their development.

PD on learning trajectories presents significant challenges. The study described here demonstrates that teachers need extended time to understand and to use learning trajectories in their practice. Even with 20 hours, activities grounded in their daily practices, and sustained support over the

14 weeks, the effects of the ERNR PD were modest. Additionally, the culture of schooling reinforces a view of student understanding in terms of grade levels and labels of proficiency assigned by high-stakes testing, whereas a perspective on learning based on learning trajectories promotes a broader view of student learning that is more related to prior experiences and opportunities to learn.

One last challenge is of particular importance when considering PD on learning trajectories—the ways that learning trajectories interact with existing curricula. Because there is considerable variation in the ways curriculum develops mathematical understanding (Stein, Remillard, & Smith, 2007), there is a significant risk of misalignment between existing curricula and learning trajectories. For example, most curricula do not provide explicit attention to the idea of equipartitioning, leaving teachers to make adaptations and choices to incorporate equipartitioning in the early grades. Therefore, PD on learning trajectories should provide ample time for and support in identifying curricular opportunities as well as understanding the potential interactions and effects of such adaptations.

AUTHOR'S NOTE

The work reported in this chapter is a part of the author's dissertation study, *Teachers' Uses of Learning Trajectory for Equipartitioning*, completed at North Carolina State University under the direction of Dr. Jere Confrey. This work was supported with funding from the National Science Foundation (DRL-073272) for the Diagnostic Learning Trajectories Approach (DELTA) project. Views expressed here are those of the author only, and do not necessarily represent the views of the National Science Foundation. I wish to thank Jere Confrey for her guidance, feedback, and encouragement and under whose direction this study was conducted. Additionally, I express gratitude to members of the Diagnostic eLearning Trajectories Approach (DELTA) research group at North Carolina State University for feedback, assistance, and support: Alan Maloney, Gemma Mojica, Kenny Nguyen, Marrielle Meyers, Ryan Pescosolido, Cyndi Edgington, Ayanna Franklin, and Zuhal Yilmaz.

REFERENCES

Battista, M. (2004). Applying cognition-based assessment to elementary school students' development of understanding of area and volume measurement. *Mathematical Thinking and Learning, 6*(2), 185–204.

Black, M. (1962). *Models and metaphors: Studies in language and philosophy*. Ithaca, NY: Cornell University Press.

Clements, D., Wilson, D., & Sarama, J. (2004). Young children's composition of geometric figures: A learning trajectory. *Mathematical Thinking and Learning, 6(2)*, 163–184.

Cobb, P., & Steffe, L. P. (1983). The constructivist researcher as teacher and model builder. *Journal for Research in Mathematics Education 14*(2), 83–94.

Cobb, P., Confrey, J., diSessa, A., Lehrer, R., & Schauble, L. (2003). Design experiments in educational research. *Educational Researcher, 32*(1), 9–13.

Cohen, D., & Hill, H. (2001). *Learning policy: When state education reform works.* New Haven, CT: Yale University Press.

Confrey, J. (2006). The evolution of design studies as methodology. In K. Sawyer (Ed.), *The Cambridge handbook of the learning sciences* (pp.131–151). New York, NY: Cambridge University Press.

Confrey, J., & Lachance, A. (2000). Transformative teaching experiments through conjecture-driven research design. In A. Kelly & R. Lesh (Eds.), *Handbook of research design in mathematics and science education* (pp. 231–265). Mahwah, NJ: Lawrence Erlbaum Associates.

Confrey, J., & Maloney, A. P. (2012). A next generation digital classroom assessment based on learning trajectories. In C. Dede & J. Richards (Eds.), *Digital teaching platforms* (pp. 134–152). New York, NY: Columbia University Press.

Furtak, E. M. (2009, June). *Toward learning progressions as teacher development tools.* Paper presented at the Learning Progressions in Science (LeaPS) Conference, Iowa City, IA. Retrieved from http://www.education.uiowa.edu/projects/leaps/proceedings/Default.aspx

Glaser, B. (1992). *Basics of grounded theory analysis.* Mill Valley, CA. Sociology Press.

Mojica, G. (2010). *Preparing pre-service elementary teachers to teach mathematics with learning trajectories.* Unpublished doctoral dissertation, North Carolina State University, Raleigh, NC.

Mojica, G. F., & Confrey, J. (2009). Pre-service elementary teachers' utilization of an equipartitioning learning trajectory to build models of student thinking. In M. Tzekaki, M. Kaldrimidou, & H. Sakonidis (Eds.), *Proceedings of the 33rd conference of the International Group for the Psychology of Mathematics Education* (Vol. 4, pp. 129–136). Thessaloniki, Greece: International Group for the Psychology of Mathematics Education.

Powell, A. B., Francisco, J. M., & Maher, C. A. (2003). An analytic model for studying the development of learners' mathematical ideas and reasoning using videotape data. *Journal of Mathematical Behavior, 22*(4), 405–435.

Stein, M., Remillard, J., & Smith, M. (2007). How curriculum affects student learning. In F. Lester (Ed.), *Second handbook of research on mathematics teaching and learning* (pp. 319–370). Charlotte, NC: Information Age Publishing.

TERC. (2007). *Investigations in number, data, and space: Second edition.* Glenview, IL: Pearson/Scott Foresman.

Wilson, P. H., Lee, H. S., & Hollebrands, K. F. (2011). Understanding prospective mathematics teachers' processes for making sense of students' work with technology. *Journal for Research in Mathematics Education, 42*(1), 39–64.

CONCLUSION

LEARNING TRAJECTORIES GOING FORWARD

A Foundation for Coherence in the Instructional Core

Jere Confrey, Alan P. Maloney, and Kenny H. Nguyen

ROOTS OF LEARNING TRAJECTORY RESEARCH

Piaget's foundational research established that children's points of view do not mirror those of adults, nor are they merely imperfect or partial copies of adults' ideas. In creating the origins of constructivism, Piaget offered three key constructs that still influence much of the work on learning trajectories: *genetic epistemology*, *schemes*, and *stages*.

With genetic epistemology, Piaget (1970) recognized that it is not only *what* one asserts that constitutes one's knowledge, but instead it is *how* and *why* one comes to believe an idea that determines the idea's worth. From Piaget's theories, the learning trajectories researcher is compelled to ask the questions: How did a particular idea originally evolve? What outstanding problems did it resolve? And what kinds of opportunities to make sense

Learning Over Time: Learning Trajectories in Mathematics Education, pages 243–255.

of the world did it engender? Asking these questions of a "big idea" in mathematics learning gives one an opportunity to consider a mathematical idea from the perspective of the child—not just what resources and beliefs a child brings to the situation, but what the new idea permits a child to do that she or he might want to do, and how this changes our own understanding of more authorized expert knowledge. Confrey (1995, 1998) referred to this challenge as negotiating the "voice-perspective" dialectic, arguing that carefully listening to the "voice" of a learner interacts with the "perspective" of the listener (the expert—a teacher or researcher), and triggers changes in the listener's own perspective, or the listener's understanding of his or her own knowledge. Genetic epistemology helps the researcher to consider the underlying "problematic" and "create a need for the idea" (Confrey, 1991).

The second construct from Piaget, the scheme, proves useful to the learning trajectories researcher because it places a concept within an action-oriented perspective that is also anticipatory. A scheme evolves when a way of solving a problem proves itself to be practical, satisfying, and reasonably efficient. A scheme is a way of anticipating, recognizing, and then attempting to apply that same frame to solve an unfamiliar problem (Piaget & Inhelder, 1969). Schemes help to explain the movement forward in a learning trajectory, and they help to explain why the sequencing of tasks is so critical to learning.

The third construct from Piaget (and other developmental psychologists) that is often erroneously tied to learning trajectories is that of stages (i.e., stages of development). For Piaget, stages were, first of all, broad conceptions of generic approaches of how children operated in relation to knowledge forms. They expressed constraints on how children perceive, and they implied or expressed limits to children's understanding, rooted in biological/psychological developmental patterns. They also framed how children made sense of the world in relation to their interactions with materials and operations used across a variety of topics. Furthermore, stage theory was viewed as a limiting factor: Until a child moved to a new stage, certain kinds of thinking and operations were said to be impossible in the current stage. Finally, stage theory was soundly criticized for implying universality—descriptions of absolutes that underlay the biology and physiology of the brain and body. Research has suggested that many such broad and universal descriptions of stages can be challenged.

Some researchers have called into question the validity of learning trajectories because an association with stage theory seemed to them to imply that trajectories are completely linear, inflexible paths and not conducive to developing critical thinking in students (Lesh & Yoon, 2004). Learning trajectories certainly rely on a sequencing of activities or tasks to support children's conceptual and developmental learning; often refer to states or transitions in knowledge; and take into account, where it seems necessary, aspects of the biological development of young children (see Clements & Sarama, Chapter 1, this volume) that might constrain children's abilities to accomplish certain tasks or reasoning. It is also the case that novice learners appear to be more likely to profit from certain kinds of inscriptions, materials, tasks, and tools than from others, and that these depend on combinations of context, instruction, prior experience, and biological-psychological development. But these features and qualities—and the commitments of learning trajectory research—differ drastically from what is implied by stage theory.

Learning trajectory researchers' goals are to discover the ideas, concepts, and skills that children can express, articulate, and develop when they are given the opportunity to engage with rich and interesting activities and to express and reconsider their experience and understanding. Learning trajectory researchers share a commitment to careful sequencing of student activities, regarding this as necessary to support development (and which incidentally is a rejection of unstructured discovery learning). Although learning trajectories are often expressed or represented as sequences of knowledge states, this does not imply a strict linear or ladder-like organization of the cognitive levels within learning trajectories, but rather it acknowledges the need to identify tasks and activities that can be undertaken by students in an order that appears to fruitfully support growth in sophistication in students' reasoning and cognition.

Learning trajectory researchers need to understand the relevant content from an expert's perspective, to ensure that the learning trajectory supports eventual accomplishment of the overall learning goals of the trajectory. It does not follow from this, however, that learning trajectories should be comprised of sequences of prerequisites for individual concepts. This was an approach to curriculum advocated by Gagné (1971), a classic curriculum theorist, who is known for expressing the question, "What is needed to know in order to learn *x*?" and who proposed a systematic process to identify logical prerequisites. Arguing that instruction should follow a pure mathematical progression based on the development of mathematical structures does not do justice to the way that children approach the learning of mathematical topics. While such back-mapping can be helpful at the beginning of proposing a learning trajectory, it is only a logical thought experiment and should not be confused with the empirical research that is necessary to understand the genesis of children's understanding and reasoning (Confrey, Maloney, Wilson, & Nguyen, 2010).

Perhaps the primary commitment shared by all the authors in the first section of this volume is that listening to children is a key element of successful instruction in mathematics. The kind of listening envisioned by the authors is not a skill that comes naturally or spontaneously to most people, not even to most teachers. Just as a piece of music can be listened to at some level by anyone, what novices hear and what experts hear differs considerably. While a novice may experience pleasure or excitement, novices do not typically recognize all the elements of harmony, rhythm, melody, and dynamics in a complex piece. Similarly, listening to children requires considerable training to hear or see and interpret the nuances of children's actions, utterances, and choices, as children undertake carefully designed tasks and struggle to make sense of them. Children create inscriptions and discuss their ideas with others, propose and explain ways to solve their tasks or check their work, and engage in self-reflection. These authors know and have shown that this kind of listening can be learned through practice, examples, and the articulation of what has already been shown through careful empirical study. They also know that listening to children is a generative skill: As one develops more expertise in listening, opportunities increase to hear new perspectives and to gain additional insight into children's learning.

A second commitment shared by the authors is that the process of learning involves epistemic acts—acts that promote the generation, revision, and assertion or justification of knowledge. Epistemology, by philosophical tradition the study of the nature of knowledge, was traditionally defined by three attributes: truth, belief, and justification, or simply as "justified true belief" (Confrey, 1999, p. 9). The tradition of epistemology was challenged by philosophers of science who questioned whether knowledge could be ultimately ordained as universal, absolute and immutable at a moment in time or a point in space—or if it had to be studied and understood in situ and as an interaction between the knower and the known. For many, knowledge is situated: (1) in context, in relation to culture and tools; (2) in structural relationships, connected to other ideas; and (3) in time, evolving with the solution to a particular problem or revealed through a particular process. Philosophers of science have also documented that knowledge can shift relatively abruptly, displacing what has gone before when confronted by a new alternative conception or by a situation that stimulates reframing or reorganizing what one believes one knows. Only by understanding the movement and evolution of ideas can one thereby understand the basis for the assertion of knowledge.

The researchers in this volume all acknowledge that learning trajectories remain, to a greater or lesser extent, conjectures or hypotheses about progressions of states of student knowledge. This demonstrates the scientific nature of learning trajectory research—an empirically grounded investigation of student knowledge growth, subject to conjecture, modeling, and hypothesis testing, aimed at generating more predictive understanding of student knowledge and reasoning growth in a context of student engagement with rich and carefully developed tasks. "The researchers in this volume employ the methodologies of learning trajectory research to make the best available characterization of progression of student knowledge of big ideas (concepts) within the relevant domains.

The learning trajectories being developed today, rooted as they are in empirical research, are intended by the authors to be much more broadly reliable as characterizations of progression of student learning than individual teachers' hypotheses (in the original sense of Simon's (1995) definition) for day-to-day teaching. Thus, a third commitment shared by the researchers in this volume is to stabilize learning trajectories sufficiently so they can be deployed and adapted as a broadly shared basis for classroom instruction, curriculum, assessment, and/or policy and standards—in effect, as theories of learning for mathematical concepts. This work has really just begun; there are many concepts—many big ideas in K–12 mathematics—for which research to develop learning trajectories has not even begun.

THE GRAIN SIZE OF LEARNING TRAJECTORIES

With these broad strokes regarding the origins of learning trajectory research, what can be said about them and what they contain? Fundamentally, there are three main features or components: (1) the prior knowledge that students bring to the classroom, (2) a domain goal of understanding, and (3) a series of states of knowledge that map a path from the prior knowledge to the goal. The domain goal specifies knowledge and reasoning that are important to a particular concept (or group of related concepts) in the discipline of mathematics. Each of these three features poses its own challenges, as will be described below.

Learning theorists acknowledge the importance of prior knowledge. Learning trajectories researchers enter with an understanding that it is not a straightforward process to find out what this prior knowledge is—the learners themselves typically do not know what the related ideas are and how to call them up (one of the challenges of establishing the progression of states of knowledge, and, incidentally, one of the correlate goals of implementing instruction based on learning trajectories). The researchers must attempt to join the setting with an open, investigative, and curious mind, recognizing that understanding a student's initial perspective is going to be a gradual process and will keep shifting as different elements of past experience and current expectations are elicited and when they work with students groups of diverse backgrounds.

Even the instructional target identified by the learning trajectory researcher may have mutable qualities. For most researchers, the "big idea" or "domain goal" is viewed as "established," a concrete endpoint for what is to be learned. However, nearly all researchers report that after studying children's progress toward that goal, their own perspective of that target shifts, imbuing it with more differentiated meanings even while retaining its power as a compelling idea (Corcoran, Mosher, & Daro, 2011). A big idea is one with strong explanatory power, fertile connections, and, usually, a long runway of learning. For some researchers, the target itself can be negotiated over the course of research and instruction, although to retain a role in or a link to a learning trajectory, that change in target should be carefully and compellingly explained as a product of the instructional interactions.

Grain sizes vary for different learning trajectories. So too, therefore, does the anticipated duration of instruction to achieve an intended domain goal of understanding, the ages of the learners across the relevant population, the level of detail in the learning trajectory, and the number of states or levels. These differences have significant implications for what is studied and how it is investigated. In this volume, the grain sizes of the overall trajectories range from months to years, and descriptions of the levels or states are variously described as six plateaus (Barrett & Battista, Chapter 4, this volume), to 16 proficiency levels (Confrey et al., Chapter 3, this volume), to seven constructs with multiple performance levels in each (Lehrer et al., Chapter 2, this volume). The time span over which a learning trajectory's complete span of proficiencies can be accomplished is neither predetermined nor necessarily obvious. For instance, neither the equipartitioning (Confrey et al., this volume) nor the data modeling (Lehrer et al., this volume) learning trajectory is currently embedded in most curricula; it would appear that the time to achieve the top proficiency levels is highly contingent on curriculum and instructional approaches in a particular situation. Confrey and colleagues constructed the equipartitioning learning trajectory with the assumption that the highest proficiency levels would not be accomplished until grade 5 or even later, but they envision that it might be possible for students to accomplish the equipartitioning proficiency levels in as little as two years (grades 2 through 3), given sufficient teacher knowledge and explicit curricular treatment of fair sharing and its consequences in the curriculum. The studies by Lehrer et al. on the data modeling and statistical reasoning learning progression demonstrated highly productive growth in student reasoning on these topics over the span of grades 5 and 6; incorporating their learning trajectory/progression in schools could conceivably lead to accelerated student accomplishment of the domain learning goals.

INTERMEDIATE STATES OF LEARNING TRAJECTORIES

The heart of the empirical research on learning trajectories is characterizing the states of knowledge that represent the transitions from prior knowledge to goal understanding. The authors in this book recognize that we need better descriptions of how learners progress from their own prior knowledge to more sophisticated reasoning and more advanced competencies.

Learning trajectories researchers recognize the critical role of tools as mediational artifacts, and so also draw heavily from *sociocultural perspectives.* Learning does not just happen based on inherent processes of biological or psychological maturation—learning is fostered by appropriate kinds of experience; rich tasks and tools; and supportive kinds of interactions and opportunities for exchange, critique, mediation, and reflection. Learning is shaped by interactions with others—whether in direct interactions with a peer or a more knowledgeable teacher or other person, or more indirectly in relation to cultural norms, media, material, and tools.

The researchers represented in this volume understand that instruction is critically important to students' progress through learning trajectories, and they express different views of how to recognize and account for the central role of instruction in relation to a learning trajectory. The originality of Lehrer et al.'s approach, and the robustness of the sequence of activities that support growth of student reasoning it has identified are in large part due to its being situated within a long-term teaching experiment. Lehrer et al. hew rather closely to Simon's (1995) conception of learning trajectory as conjectured states of student knowledge along with the means of supporting transitions between them, while defining the learning progression as the learning trajectory *along with* the necessary related professional development, assessment, and institutional organization (Lehrer et al., Chapter 2, this volume). Clements, Sarama, and Barrett (Chapter 1) have embedded their learning trajectory for early number and operation reasoning within curricular treatments, distinguishing it from developmental paths that might be assumed to exist external to instruction. Battista (2004, and Barrett & Battista, Chapter 4, this volume) does not see a single trajectory running through a domain but rather a cognitive terrain consisting of several plateaus of knowledge from informal conceptualization and reasoning to formal mathematical reasoning with several possible learning trajectories weaving through the plateaus. He calls this a levels model for a topic, with such a model including the cognitive terrain in which the learning trajectories exist. In contrast, while Confrey et al. (Chapter 3) recognize that careful curriculum design and instructional moves are required to create the conditions for learning, they emphasize big ideas and conceptual understanding embedded in standards and across curricula and seek to define learning trajectories to have a degree of independence from any particular curricular treatment. Their focus is to eventually create diagnostic assessments, based on such learning trajectories, that can be broadly deployed to ascertain students' understanding of big ideas in mathematics and thereby to offer teachers structured and coherent feedback on students' conceptual growth regardless of the curriculum the teachers might be using. They also acknowledge the key role of professional development focused on the learning trajectories in improving teachers' understanding and use of concept-based instruction and ability to eventually make optimal use of feedback developed from such learning trajectory-based diagnostic assessment.

Ultimately, all the authors represented in the first half of this volume are working toward a means of describing some of the most essential and typical intermediate states during learning, ways for teachers and researchers to recognize transitions, and, in many cases, development of assessments and other tools to support the detection of student growth and to identify effectively the particular cognitive proficiencies exhibited, and the setbacks or misconceptions encountered, by students. The various authors all identify levels or knowledge states with their respective learning trajectories/progressions (named variously "cognitive levels," "proficiency levels," "performance levels," "constructs"). Several qualities of this part of learning trajectory work can be summarized, as below.

Determining when a set of behaviors should be described as a distinct state is a major challenge of this work and depends on multiple lines of evidence, iteratively considered and refined. The iterative nature of this type of research is one of its features that suggests the wisdom of characterizing learning trajectories as conjectured and hypothetical rather than final or fixed, even as we attempt to stabilize them sufficiently for use across multiple venues (in classrooms locally, broadly in professional development and curriculum development, and as the basis of standards and assessment tools). Ultimately, it is essential to characterize each of these states in enough detail so that they are recognizable by teachers and researchers who observe the students solving tasks, review student work, guide discourse, or interview them.

In general, the learning trajectories described in this book are grounded in a flexible view of sequencing. The orderings of states or concepts described in the learning trajectories are considered to be typical and expected, but are also contingent on instruction, discourse, tools, and representations, and student inscriptions. Levels within learning trajectories are intended to be distinct, but are not envisioned as dependent strictly on the previous level. Levels are influenced by combinations of earlier accomplishments, and it is assumed that features or reasoning first recognized as characteristic of previous levels are subsequently called on by students to facilitate further progress. (See also the detailed discussion of levels and progression through a learning trajectory in Chapter 1 by Clements and Sarama, this volume.)

Another challenge experienced by researchers is to specify a state that is "partially correct" and "partially incorrect," or "limited." These are intended simply as descriptions of what students know and believe. It is conceivable that some of the identified proficiency levels could at first glance be regarded on occasion as inappropriately reinforcing errors or misconceptions or limiting growth by delineating these categories. The learning trajectory researcher's response is to emphasize that these are descriptive expected tendencies or likely probabilities of descriptive states that commonly occur, based on the scientific research into the students' actual learning. How these states are treated in instruction and assessment is then subject to further discussion, conjecture, and empirical study.

A third quality is to consider what kinds of constructs comprise the states. Is it the understanding of a property, a concept, a strategy, or a relationship? In these chapters, the definitions or constructs of learning trajectories vary but seem to share a quality that once a child has gained a new proficiency, the way s/he perceives the type of task changes. Some of these proficiencies may be regarded as transformative, propelling the students' reasoning and reorganization of their own knowledge. Such states or proficiency levels could be described in terms of what they permit students to approach from a different vantage point.

All of these considerations that are particularly germane to the states or levels that make up the central portions of learning trajectories and the transitions among them—the distinctions among states, the types of constructs that define them, sequencing of states, and the role of states that comprise partially correct knowledge—can be referred to as the messy middle (Confrey et al., 2010). One reason for referring to this as the "messy middle" is to indicate that the structure and sequencing of the states in the middle of the learning trajectory varies by researcher (and therefore also by the particular learning trajectory). For some, these may be regarded as conceptual prerequisites—an earlier state is assumed to be required for successful transition to a later state. For others, the middle states are a collection of topics that increase the likelihood of achieving deep understanding at the target, but it may be much less clear whether the entire collection of topics is both necessary and sufficient for achieving the goal level of understanding, even if the states may appear to be cognitively distinct. And for other researchers, it is conjectured that students will move around in this space—holding some constructs at some points in time, dropping back under cognitive pressure when encountering unfamiliar tasks, and then moving on at others.

A final reason to acknowledge and anticipate the messiness of the middle is that student behavior in relation to tasks characteristic of the broad middle states of a learning trajectory will be influenced by students' instructional experience. The extent to which students' paths through the states of a trajectory is influenced (and this may include reordering states or even skipping some states altogether) by experiences that occur parallel to or interspersed among the epistemic activities directly germane to a particular learning trajectory is not generally understood. Whether students go through cycles of advancing and dropping back within a learning trajectory is also a critical question in learning trajectory research (see, in particular, Chapters 1 and 4, this volume). This aspect of the research, especially, for large grain-size learning trajectory and those more advanced domain goals in higher grades, will involve the milieu of topics encountered as students proceed up the grades, via different curricula and with access to different resources. Over the long term, researchers will have to conduct more studies that reflect the actual instructional practice—providing curriculum, tools, and professional development for instruction in order to characterize learning trajectories within student learning environments. If the learning trajectory construct is expected to form and mature across grades, the challenges connected with describing the middle phases are critical to address in the research. We note finally that the flexibility of or variability in individual students' paths through the trajectory, including possible skipping states, or cycling through various proficiency levels is in tension with representations of learning trajectories as linear or two-dimensional graphical displays.

LEVERS FOR POLICY AND PROFESSIONAL DEVELOPMENT

Multiple authors acknowledge the value of learning trajectories as constructs that promote coherence in the practice of mathematics instruction. Some of the enthusiasm for learning trajectories stems from their potential to function as boundary objects that can promote collaboration for systemic improvement, across communities of researchers, teachers, curriculum writers, and professional developers.

The authors in the second section of the volume illustrate various ways that such coherence might be promoted through the use of these boundary objects (with the definition of learning trajectory varying from author to author, as one might expect to be the case for a boundary object). The discussion of the role of learning trajectories in standards shows how the research community has sought to incorporate lessons learned from student learning research into standards, by synthesizing research results so that they can guide the development of standards. The development of the Common Core State Standards in Mathematics, or CCSS-M (National Governors Association Center for Best Practices, Council of Chief State School Officers, 2010) with learning progressions as one core element (along with international benchmarking, and a sharpening of focus) demonstrates the possibilities of the positive role for learning trajectories at the policy level. Implementation of the learning trajectories in CCSS-M is incomplete, due in part to compromises that had to be struck during writing and to the incomplete coverage in the research base itself, so that consistency and coherence of the standards themselves in relation to learning trajectories was only partially realized in the document published in 2010. However, the stated intentions of the Common Core State Standards Initiative that the standards be improved over time as additional research is brought to bear on the topics, offers the hope of iterative ongoing improvement with respect to these issues.

Likewise, the development of ways to conduct a learning trajectories analysis of existing curriculum demonstrates that learning trajectories can provide a powerful addition to our ability to conduct content analyses, one critical means of evaluating curricular effectiveness (National Research Council, 2004). Nguyen and Confrey's chapter and Olsen's chapter illustrate the potential of highlighting distinctions, within and among curricula, that could then later be linked to differences in student outcome. Those two chapters presage what is likely to become more prevalent in the future, the use of learning trajectories in the design of curricular programs.

While not addressed directly in the volume, the topic of assessment is another policy lever to be potentially influenced through the development of learning trajectories. Already, we see such discussions emerging in relation to both formative uses (Heritage, 2008, 2010), diagnostic uses of learning trajectories (Confrey, Maloney, Nguyen, & Corley, 2012), and summative uses (PARCC, 2012; Smarter Balanced Assessment Consortium, 2012).

Learning trajectories may also provide another kind of lever improvement of instruction, when used as the framework for professional development programs, with strong potential for improving teacher understanding of student learning and promoting cross-grade discourse and content knowledge. The chapter by Wilson illustrates the use of a learning trajectory as the foundation of professional development with in-service elementary teachers. In this work, Wilson demonstrates that as teachers learn to make effective use of learning trajectories, they will experience changes in how they build models of student thinking: Their own conceptual perspectives will change. In addition, they will engage in changing a number of their activity structures, such as how they present and manage tasks, how they support discourse, and how they assess children's knowledge on an ongoing basis. This potential has been further articulated recently by Sztajn, Confrey, Wilson, and Edgington (2012).

Learning trajectories as boundary objects signal a step forward for mathematics education—a means of framing coherent, mutually supportive efforts in mathematics education research and instruction. Rather than aiming for progress in a loosely coupled system in which the only clear policy levers are the bookends of standards and assessment, a learning trajectories approach can provide a means to stabilize and focus both the accountability system and the instructional core around a fundamental basis in the learning sciences.

GOING FORWARD

The authors in this volume address critical ideas in learning trajectory development for mathematics education and their potential applications. They demonstrate how the work builds on both constructivist and sociocultural perspectives to define how big ideas develop over time. All of these researchers acknowledge the conjectural nature of learning trajectories while seeking to establish a degree of stability of particular learning trajectories, operationalized for more broad use within the instructional core of daily classroom learning. All are dedicated to further empirical research to develop additional learning trajectories and to study the validity, effectiveness, and advantages offered by the use of learning trajectories in instruction, assessment, curriculum, and professional development.

REFERENCES

Battista, M. T. (2004). Applying cognition-based assessment to elementary school students' development of understanding of area and volume measurement. *Mathematical Thinking and Learning, 6*(2), 185–204.

Confrey, J. (1991). Learning to listen: A student's understanding of powers of ten. In E. von Glasersfeld (Ed.), *Radical constructivism in mathematics education* (pp. 111–138). Dordrecht, Netherlands: Kluwer Academic Publishers.

Confrey, J. (1995). Student voice in examining "splitting" as an approach to ratio, proportions and fractions. In L. Meira & D. Carraher (Eds.), *Proceedings of the 19th Annual Meeting of the International Group for the Psychology of Mathematics Education* (Vol. 1, pp. 3–29). Recife, Brazil: Universidade Federal de Pernambuco.

Confrey, J. (1998). Voice and perspective: Hearing epistemological innovation in students' words. In M. Larochelle, N. Bednarz, & J. Garrison (Eds.), *Constructivism and education* (pp. 104–120). New York, NY: Cambridge University Press.

Confrey, J. (1999). Voice, perspective, bias and stance: Applying and modifying Piagetian theory in mathematics education. In L. Burton (Ed.), *Learning mathematics: From hierarchies to networks* (pp. 3–16). New York, NY: Falmer Press.

Confrey, J., Maloney, A. P., Nguyen, K. H., & Corley, A. K. (2012, April). *A design study of a wireless interactive diagnostic system.* Paper presented at the annual meeting of the American Education Research Association, Vancouver, British Columbia.

Confrey, J., Maloney, A. P., Wilson, P. H., & Nguyen, K. H. (2010, May). *Understanding over time: The cognitive underpinnings of learning trajectories.* Paper presented at the Annual Meeting of the American Education Research Association, Denver, CO.

Corcoran, T., Mosher, F. A., & Daro, P. (Eds.). (2011). *Learning trajectories and progressions in mathematics.* New York, NY: Consortium for Policy Research in Education.

Gagné, R. M. (1971). *The conditions of learning.* New York, NY: Holt, Rinehart & Winston.

Heritage, M. (2008). *Learning progressions: Supporting instruction and formative assessment.* Washington, DC: Chief Council of State School Officers.

Heritage, M. (2010). *Formative assessment and next-generation assessment systems: Are we losing an opportunity?* Paper prepared for the Council of Chief State School Officers. Los Angeles, CA.

Lesh, R., & Yoon, C. (2004). Evolving communities of mind—In which development involves several interacting and simultaneously developing strands. *Mathematical Thinking and Learning, 6*(2), 205–226.

National Governors Association Center for Best Practices & Council of Chief State School Officers. (2010). *Common core state standards for mathematics.* Washington, DC: Authors. Retrieved from http://www.corestandards.org/assets/CCSSI_Math Standards.pdf

National Research Council. (2004). *On evaluating curricular effectiveness: Judging the quality of K–12 mathematics evaluations.* Washington, DC: The National Academies Press.

PARCC. (2012). Partnerhsip for assessment of readiness for college and careers. Retrieved from http://www.parcconline.org

Piaget, J. (1970). *Genetic epistemology.* New York, NY: W.W. Norton.

Piaget, J., & Inhelder, B. (1969). *The psychology of the child.* New York, NY: Basic Books.

Simon, M. A. (1995). Reconstructing mathematics pedagogy from a constructivist perspective. *Journal for Research in Mathematics Education, 26*(2), 114–145.

Smarter Balanced Assessment Consortium. (2012). Overview Presentation. Retrieved from http://www.smarterbalanced.org/wordpress/wp-content/uploads/2011/12/Smarter-Balanced-Overview-Presentation.pdf

Sztajn, P., Confrey, J., Wilson, P. H., & Edgington, C. (2012). Learning trajectory based instruction: Toward a theory of teaching. *Educational Researcher, 41*(5), 147–156.

Comp: Please note, author bios still pending from Ayers, Kim, Lehrer, and M. Wilson (Chapter 2).

AUTHOR BIOGRAPHIES

Elizabeth Ayers is a Psychometrician-Statistician working at American Institutes for Research (AIR) in Washington, DC. She obtained her PhD in statistics from Carnegie Mellon University and was a post-doctoral fellow in the Graduate School of Education at the University of California, Berkeley. During her post-doc she managed the IES funded Assessing Data Modeling and Statistical Reasoning project which sparked her interest in learning trajectories. In addition, Dr. Ayers conducts research and publishes in the area of cognitive diagnosis models.

Jeffrey E. Barrett is professor of mathematics education at Illinois State University and serves as an associate director of the Center for Mathematics, Science and Technology at Illinois State University. His primary research interests include the learning and teaching of the mathematics of measurement and geometry, the use of computer software to model mathematical ideas, and the professional development of teachers engaged in teaching mathematics and measurement-related science. Dr. Barrett directed the four-year project, A Longitudinal Examination of Children's Developing Knowledge of Measurement: Mathematical and Scientific Concept

Learning Over Time: Learning Trajectories in Mathematics Education, pages 257–264.

and Strategy Growth from Pre-K to Grade 5 in collaboration with Douglas Clements and Julie Sarama at the University at Buffalo, SUNY between 2007 and 2011, a project funded by the DR K–12 and REESE Programs of the National Science Foundation. Barrett has studied learning trajectories as tools for examining the task-based reasoning of students as they develop length, area and volume measurement concepts. Using teaching experiment methods, his work has addressed the integration of schemes for number and space as conceptual bases for understanding spatial measurement in elementary school.

More recently, Barrett has demonstrated the relevance of learning trajectory accounts of spatial measurement as a basis for teacher development. His ongoing work with elementary teachers is an adaptation of lesson study processes to focus on mathematical content as a theme, especially measurement concept growth. Barrett examines the professional growth of teachers by providing teachers opportunity to frame their own questions about student thinking, to collect evidence of students' expressions of thinking and reasoning, and to check their own hypotheses about student learning.

Michael T. Battista is professor of mathematics education in the School of Teaching and Learning at Ohio State University. He has taught mathematics to students of all ages, from preschool through adult, and has been involved in mathematics teacher education both at the preservice and in-service levels. His major research interest is students' learning of mathematics, with most of his research focusing on students' learning of geometry and geometric measurement and the use of technology in mathematics teaching. He has been conducting research on learning progressions in mathematics for over 25 years.

Battista is currently completing a two-phase NSF-funded project on cognition-based assessment (CBA) materials for use in elementary school mathematics. These materials will be published by Heinemann and will cover operations on whole numbers, place value, fractions, geometric measurement, and geometric shapes. In Phase 1, he developed and tested CBA materials, which have three components: (1) Descriptions of core ideas and reasoning processes in elementary school mathematics; (2) for each core idea, theoretical frameworks that describe the cognitive processes, milestones, and developmental landscapes for the idea; and (3) for each core idea, assessment items that reveal students' cognitions and precisely locate students' position in learning trajectories for the idea. In Phase 2, he has been studying how elementary teachers learn and make pedagogical sense of research on mathematics learning as depicted in CBA materials. This phase of the project is investigating: (1) teachers' understandings of students' mathematical thinking before and after instruction on that thinking, (2) the processes by which teachers learn about students' mathematical thinking while participating in instruction on that thinking, (3) factors affecting teachers' learning of this material, and (4) effects of learning this material on teachers' conceptualizations of mathematics learning, teaching, and assessment.

Battista has begun an NSF-funded project to develop and test an individualized computer-based dynamic geometry curriculum for elementary and middle school that integrates his work on learning progressions in geometry and measurement, formative assessment, and dynamic geometry instruction.

Douglas H. Clements is Kennedy Endowed Chair in early childhood learning and professor at the University of Denver. Previously a kindergarten teacher for five years and a preschool teacher for one year, he has conducted research and published widely in the areas of the learning and teaching of early mathematics and computer applications in mathematics education. His most recent interests are in creating, using, and evaluating a research-based curriculum and in taking successful curricula to scale using technologies and learning trajectories. He has published over 120 refereed research studies, 18 books, 70 chapters, and 275 additional publications. His latest books detail research-based learning trajectories in early mathematics education: *Early Childhood Mathematics Education Research: Learning Trajectories for Young Children* and a companion book, *Learning and Teaching Early Math: The Learning Trajectories Approach* (Routledge).

Dr. Clements has directed 20 projects funded by the National Science Foundation (NSF) and the U.S. Dept. of Education, Institute of Education Sciences (IES). Currently, he is principal investigator on two large-scale randomized cluster trial projects (IES). He is also working with colleagues to study and refine learning trajectories in measurement (NSF). Two recent research projects have just been funded by the NSF. Clements is PI on the first, *Using Rule Space and Poset-based Adaptive Testing Methodologies to Identify Ability Patterns in Early Mathematics and Create a Comprehensive Mathematics Ability Test,* which will develop a computer-adaptive assessment for early mathematics. He is co-PI on the second, *Early Childhood Education in the Context of Mathematics, Science, and Literacy,* developing an interdisciplinary preschool curriculum. Clements was a member of President Bush's National Math Advisory Panel, convened to advise the administration on the best use of scientifically based research to advance the teaching and learning of mathematics and coauthor of the Panel's report. He was also a member of the National Research Council's committee on early mathematics and co-author of their report. He is presently serving on the Common Core State Standards committee of the National Governor's Association and the Council of Chief State School Officers, helping to write national academic standards and the learning trajectories that underlie them. He is one of the authors of NCTM's *Principles and Standards in School Mathematics* and *Curriculum Focal Points.*

Jere Confrey is the Joseph D. Moore Distinguished University Professor of Mathematics Education at North Carolina State University. She has conducted research on student learning of mathematics and has focused in particular on student conceptions of functions, multiplicative structures, learning of rational number concepts, and diagnostic assessment technology. She developed the splitting conjecture and has recently led the development of a learning trajectory for equipartitioning as well as a set of learning trajectories to support teachers in interpreting the Common Core State Standards for Mathematics. She has taught school at the elementary, secondary, and postsecondary levels. She chaired the NRC Committee that produced "On Evaluating Curricular Effectiveness" and was a coauthor on the NRC's "Scientific Research in Education." She co-founded the UTeach program at the University of Texas in Austin, and was a founder of the SummerMath and SummerMath for Teachers programs at Mount Holyoke College. She received a PhD and MA in mathematics education from Cornell University, and BA from Duke University.

At the present time, she is on leave from the university, serving as chief mathematics officer for Amplify Learning (formerly Wireless Generation), leading the development of a digital middle grade mathematics curriculum.

Richard A. Duschl is Waterbury Chair professor of secondary education in the College of Education, Penn State University. Twice Richard has received the JRST Award for the outstanding research article published in the *Journal of Research in Science Teaching.* From 2008 to 2011 Richard served as president of National Association for Research in Science Teaching. He was editor of the research journal *Science Education* for over a decade and editor for TC Press' Ways of Knowing in Science and Math book series. More recently, he has served as chair of the National Research Council committee that wrote the research synthesis report *Taking Science to School: Learning and Teaching Science in Grades K-8* (National Academies Press, 2007) and was part of the 2009 NAEP Science Framework planning committee. Richard currently co-chairs the earth/space sciences writing team for the Next Generation Science Standards.

Min-Joung Kim earned her PhD from Vanderbilt University's Peabody College, in the Department of Teaching, Learning and Diversity. The title of her dissertation is "Tracing the naturalization of a learning progression centered assessment system in a teacher community." She is a graduate of Seoul National University of Education (B. A. in Elementary Education) and Indiana University (M. A. in Instructional Systems). She joined the faculty of the School of Education at Louisiana State University as an assistant professor in 2013.

Richard Lehrer is Frank W. Mayborn Professor of Education at Vanderbilt University's Peabody College. A former high school science teacher, he received a Ph.D. in educational psychology and statistics from the University of New York, Albany and a B.S. in Biology from Rensselaer Polytechnic Institute. His research focuses on the design of classroom learning environments that support the growth and development of model-based reasoning in science and in mathematics. In mathematics education, he has investigated development of children's reasoning about space, measure, data, and chance when children participate in classrooms where instruction is guided by teacher knowledge of student reasoning. His interests include the role of inscription and notation in the growth of disciplinary reasoning, education as a design profession, and the development of methods and measures appropriate for the study of learning in complex systems, such as classrooms and schools. He has served as co-editor of Cognition and Instruction and on several NRC committees including *Knowing What Students Know, Systems for State Science Assessment, Understanding and Improving K-12 Engineering Education in the United States and Integrated STEM Education.* He was the 2009 recipient of the Distinguished Contributions in Applications of Psychology to Education, from the American Psychological Association and is a Fellow of the American Educational Research Association. Current research studies include experimental test of a learning progression for data and statistics, cognitively-based assessment of students' reasoning about data and chance, investigation of children's learning across STEM disciplines when they engage in engineering design, and, in concert with his colleague Leona Schauble, investigation of a learning progression oriented toward student invention and revision of models of ecological systems.

Alan Maloney is a senior research fellow at the Friday Institute for Educational Innovation at NC State University. His research interests include synthesis of research on learning trajectories, the design and implementation of learning trajectories diagnostic assessments for rational number reasoning, and the development of professional development resources for mathematics learning trajectories. Originally trained in molecular and population biology, his mathematics education work began through collaboration with Jere Confrey on the design of software for mathematics education. He is a co-designer of applications that include Graphs 'N' Glyphs and the LPPSync diagnostic assessment prototype. Maloney partnered with Confrey on development of learning trajectory-based standards for North Carolina (with a team of North Carolina K-20 educators) and in developing learning trajectories that embed the Common Core State Standards for Mathematics. Currently he is the director of the TurnOnCCMath project at NC State. He received a BS and PhD in biological sciences from Stanford University.

Kenny Huy Nguyen is an upper school mathematics teacher at Catlin Gabel School in Portland, Oregon. He currently teaches integrated Algebra II/Geometry and an elective seminar in statistics and quantitative methods while continuing to pursue questions about how elementary students learn measurement, specifically the topics of area and volume. Prior to teaching at Catlin Gabel, Dr. Nguyen was a Research Associate at the Friday Institute for Educational Innovation in Raleigh, NC where he worked in collaboration with Dr. Jere Confrey and Dr. Alan Maloney on the DR K-12 Project DELTA (Diagnostic E-Learning Trajectories Approach) which developed detailed learning trajectories for rational number reasoning (specifically equipartitioning, area and volume, and division and multiplication) and examined issues in measuring students' progress on those trajectories using wireless devices. His doctoral dissertation, under the direction of Dr. Confrey, involved examining how students used equipartitioning to develop concepts of early area measurement. Dr. Nguyen holds an S.B. in mathematics from the University of Chicago, an M.A. in Learning Technologies from the University of Michigan, and a Ph.D. in mathematics education from North Carolina State University.

Travis A. Olson is assistant professor of mathematics education in the department of teaching and learning at the University of Nevada, Las Vegas. Dr. Olson holds a bachelor's degree in mathematics and political science as well as a master's degree in mathematics. He earned his PhD in curriculum and instruction under the direction of Dr. Douglas A. Grouws at the University of Missouri, where he is an alumni fellow of the Center for the Study of Mathematics Curriculum.

Dr. Olson's current research focuses on curriculum content analysis studies employing *articulated learning trajectory* and *mathematical landscape* frameworks (at various levels, K–16), and the associated implications on students' opportunities to learn and teachers' opportunities to teach particular mathematics concepts. He is also currently investigating secondary mathematics teachers' mathematical conceptions, roots of their understandings, and abilities to model mathematical problems related to key grade 6–12 mathematical topics.

André A. Rupp is an Associate Professor in the Measurement, Statistics, and Evaluation (EDMS) program in the Human Development and Quantitative Methodology (HDQM) department at the University of Maryland. His synthesis-oriented work frequently circumscribes, deconstructs, and re-arranges the current state-of-the-art of methodological research and practice at the intersection of educational and psychological measurement, applied cognitive psychology, and the learning sciences.

His research interests center around cognitively-grounded assessment approaches and associated statistical models, which broadly fall under the umbrella terms *diagnostic measurement/cognitively diagnostic assessment and diagnostic classification models (DCMs) / cognitive diagnosis models.* To this end, he works on research around model criticism and refinement through simulation studies and applied work.

From an interdisciplinary perspective he is currently most interested in exploring how an evidence-centered design framework for assessment design, scoring, and reporting and associated use models can be put to practical use for researchers who are interested in creating integrated diagnostic assessment systems. He is particularly engaged in helping to create diagnostic feedback based on process and product data from digital learning environments, which requires the integration of tools from multivariate statistics, modern psychometrics, and educational data mining. He currently collaborates on such project, with researchers at CISCO, the University of Wisconsin-Madison, the Friday Institute for Educational Innovation at North Carolina State University, and the Education Development Center.

Julie Sarama is Kennedy Endowed Chair in innovative learning technologies and professor at the University of Denver. She conducts research on young children's development of mathematical concepts and competencies, implementation and scale-up of educational reform, professional development models and their influence on student learning, and implementation and effects of software environments (including those she has created) in mathematics classrooms. These studies have been published in more than 50 refereed articles, 4 books, 30 chapters, and 60 additional publications. She has been principal or co-principal investigator on seven projects funded by the National Science Foundation, including *Building Blocks—Foundations for Mathematical Thinking, Pre-kindergarten to Grade 2: Research-based Materials Development* and *Planning for Professional Development in Pre-School Mathematics: Meeting the Challenge of Standards 2000.* She is principal investigator on her latest NSF award, entitled *"Early Childhood Education in the Context of Mathematics, Science, and Literacy."*

Dr. Sarama is also co-directing three large-scale studies funded by the U.S. Education Department's Institute of Educational Studies (IES). The first is entitled *Scaling Up TRIAD: Teaching Early Mathematics for Understanding with Trajectories and Technologies.* The second is a longitudinal extension of that work, entitled *Longitudinal Study of a Successful Scaling Up Project: Extending TRIAD.* The third, with Dr. Sarama as principal investigator, is an efficacy study, *Increasing the Efficacy of an Early Mathematics Curriculum with*

Scaffolding Designed to Promote Self-Regulation. Dr. Sarama was previously the lead co-PI at the Buffalo site on another IES-funded project, *A Longitudinal Study of the Effects of a Pre-Kindergarten Mathematics Curriculum on Low-Income Children's Mathematical Knowledge* (IES; one of seven of a cohort of national projects conducted simultaneous local and national studies as part of the IES's *Preschool Curriculum Evaluation Research* project).

Dr. Sarama has taught secondary mathematics and computer science, gifted math at the middle school level, preschool and kindergarten mathematics enrichment classes, and mathematics methods and content courses for elementary to secondary teachers. In addition, she presently is the director of the gifted mathematics program (GMP) at the University of Buffalo, SUNY. She designed and programmed over 50 published computer programs, including her version of Logo and Logo-based software activities (*Turtle Math*™, which was awarded the *Technology & Learning* Software of the Year award, 1995, in the category "Math").

P. Holt Wilson, a former secondary mathematics teacher, is an assistant professor of mathematics education at the University of North Carolina at Greensboro. His research focuses on teachers' knowledge of student mathematical thinking and the ways that teachers may be supported in organizing mathematics instruction around students' thinking. He is currently involved in a multiyear project partnering elementary grades teachers with researchers to conceptualize a model of mathematics instruction based on learning trajectories.

Mark Wilson is Professor of Education at University of California, Berkeley. He received his PhD degree from the University of Chicago in 1984. His interests focus on measurement and applied statistics. In the past year he was elected president of the Psychometric society, and also became a member of the US National Academy of Education. He has recently published three books: one, *Constructing measures: An item response modeling approach,* is an introduction to modern measurement; the second, *Explanatory item response models: A generalized linear and nonlinear approach,* introduces an overarching framework for the statistical modeling of measurements; the third, *Towards coherence between classroom assessment and accountability* is about the relationships between large-scale assessment and classroom-level assessment. He has also recently chaired a US National Research Council committee on assessment of science achievement—*Systems for state science assessment,* and is currently co-chairing another one.